Richard Fawcett Battye

What is vital force?

A short and comprehensive sketch including vital physics

Richard Fawcett Battye

What is vital force?
A short and comprehensive sketch including vital physics

ISBN/EAN: 9783337228606

Printed in Europe, USA, Canada, Australia, Japan

Cover: Foto ©berggeist007 / pixelio.de

More available books at **www.hansebooks.com**

OR,

A SHORT AND COMPREHENSIVE SKETCH,

INCLUDING

VITAL PHYSICS, ANIMAL MORPHOLOGY, AND EPIDEMICS;

TO WHICH IS ADDED

AN APPENDIX UPON GEOLOGY:
IS THE DETRITAL THEORY OF GEOLOGY TENABLE?

BY

RICHARD FAWCETT BATTYE.

LONDON:
TRÜBNER & CO., LUDGATE HILL.
1877.
All rights reserved.

PREFACE.

The present brief Treatise is essentially a suggestive work, not a demonstrative one.

The writer can refer to no one part as being more attractive, or more important, than another. So much will depend upon the individual tastes of the reader, as to where they are directed, or as to the particular vein of thought which he has most cultivated.

The work divides itself into three parts:—(1) Vital Physics, (2) Animal Morphology, and (3) Epidemics; to which is added a short Essay upon, or rather against, the Detrital Theory of Geology—a science which records the life of a past world, but now entombed in the rocks on the surface of the earth. These either directly or indirectly connect themselves with the present, and, though dead, yet they speak to the living of ages gone by.

In place of giving a table of Contents, or a long Preface, an Introduction is given, which is an Epitome of the Text. In style and substance it is simply a Syllabus of the whole, and is intended to give, in an abrupt form, a concise Outline of the Treatise.

Originality might be considered as the primary point aimed at in this succession of short essays. But, in the matter of style, it is free from that condemnation, unless it can make some slight claim, arising from its numerous defects. But, if the work is an essentially suggestive one, it is necessarily more or less an original one; though in almost every detail anticipated in some way or other.

Belgravia, 1876.

INTRODUCTION.

VITAL PHYSICS.

In giving an Introduction to the physics of vital force, as here held, in the place of a brief analysis of the text, an attempt will be made to state plainly and in few words, what are the chief or salient points considered in the following short outline.

It is admitted that the laws which govern the inorganic world are the same which govern the organic world, and in maintaining this general government of both kingdoms the law of gravitation is rejected, as it is now held.

The reasons are:—1st, That both orbital and axoidal motion in the planets is from west to east, and that both these motions arise out of one and the same force or forces. As now held, the law of gravitation completely accounts for the orbital motion; but it leaves the axoidal motion in the cold. In astronomy the facts and observations are supplied, but not explained.

2nd: That in astronomy the repellant force, or tangental, is an initiatory or starting force, and beyond that it is no force at all, but merely the outgoings of inertia; but to obtain motion sufficient to induce both orbital and axoidal motion at one and the self-same time two *active* forces are essential, and the tangental or initiatory force is necessary to start them off or throw them out of balance, but once thrown out of balance they must ever result in a continuous cycle of motion.

3rd: The existence of force is an unknown thing without *antagonism*, and for its display resistance is essential. But in speaking of force, especially one like attraction (which can be proved from the fact that it invariably obeys the mathematical law of inversely to the square of the distance), the repellant force must be equally active with the attractive, to keep the attractive in continual process of manifestation. For an active force cannot be in continual action for ages and never bring a tangental, or, so to speak, a *negative* force into composition and resolution. It is a subject that requires so little thought in the matter of equipoise of forces, not to perceive how ridiculous it is when once fairly discussed, that a negative is keeping at bay an active force for ages, and is now as supreme as ever it was, thus making a negative quite equal to an active force, and in matters of force putting all experience and facts at utter defiance.

In treating upon two forces, the attractive and repellant, it is maintained that they have equal extension in the universe, but that their ratio of acceleration is *unequal*, that of repulsion being slightly greater in acceleration: but the attractive on the other hand, is slightly stronger, as a force, than the repellant.

Leaving the subject of axoidal and orbital motion, and the motions of planetary bodies, etc., the attention is directed from the greater to the less, or from the units of masses to attraction and repulsion between atom and atom.

1st: That both these forces as fluids permeate all atoms.

2nd: That each distinct element, as gold, silver, oxygen, hydrogen, etc., differs in the degree of permeability to these fluids, one having attraction in excess over repulsion or *vice-versâ*.

3rd: That every special element has its own individual form.

4th: That each atom of each element has, not only its own individual proportion of these two forces distributed to it; but of the elements, some have the attractive fluid more on their surfaces, and others more in the centre, and the repellant on the surface, or in the centre in the reverse order to the attractive.

By the form of atoms, the degree of forces supplied, and their order of location, all the properties of malleability, toughness, hardness, softness, brittleness, transparency, etc., found in matter are explained.

The foregoing necessarily leads to the fact that all our elements are *unequally* attracting and repelling one another; or, in other words, it gives a solution to the laws of chemical affinity, in place of that of gravitation, which latter proves that each particle attracts every other particle in an equal degree. Chemical science is a direct negative to this general law.

From the foregoing the great law of Precursion is deduced as a law which governs the entire universe. Its simple meaning is *præ* before—and *curro*, I run, or the *unequal* attraction of all units of masses to a centre; as the sun, planets, etc., as so many single masses, and known in this form by the ratio of acceleration which each possesses, as calculated from the *combined* axoidal and orbital motions; and in atoms by the laws governing chemical affinities generally.

Lastly: That attractive force in each element is always rigidly fixed and never alters; but not so the repellant. This force can be disturbed. There is a point below which it cannot be disturbed without an atom ceasing to be, which, as the universe is at present constituted, is impossible. But beyond this point of zero it is capable of the greatest amount of variation, and of localising and transplanting from one point to another with amazing power of accumulation or concentration, so as in a variety of ways to greatly counterpoise the attractive force, and to dislodge it from holding atoms, in apposition with each other. This it does as an element, or imponderable, under the forms of heat, electricity, magnetism, etc., etc.

COLOUR.

Although colours enter so largely into all things both of the organic and inorganic worlds, yet the blending of them in every variety of shade and hue, more particularly pertains to the organic kingdom. But in relation to vital force it is more a proof of exposure to air and light, and of the healthy condition of an organism or otherwise, as it is exposed to light or it is withdrawn from it, than an essential part of life. We cannot treat of light in the organic world in any different form to that in which it is viewed in the inorganic world, namely, that all the *permanent* colours in bodies are known to us by reflected light, and that it is owing to some more finely elaborated mechanical property in molecular arrangement than either the microscope, or the subtle behaviour of matter under chemical changes and affinities enables us to detect; and whether the corpuscular or emissional theory, or the more elegant theory of light, known as the undulatory, be accepted, it matters little, as in reflected light the result is pretty much the same; namely, some subtle molecular arrangement in the atoms of matter gives the various shades and hues of light known under the general term of colour.

UPON ANIMAL DIFFERENTIATION AND METAMORPHOSIS.

Leaving therefore, the subject of light and colour, which refer only to reflected light in relation to organic nature, an attempt is made to found a system of morphology and differentiation in relation to the animal kingdom, which is grounded upon the assumption of all animal tissues having in their systematic distribution *three* membranes, a serous, a mucous, and a contractile membrane, in contra-distinction to vegetables, which have only two membranes—an outer and an inner, or a serous and mucous membrane, if such nomenclature is permissible.

The term membrane is used much in the sense of rock in geology. It does not necessarily mean a continuous structure, but it refers rather to *function* running along with certain kinds of structure, and as such shows itself in a variety of forms and differentiations. Thus muscles are called *contractile membrane*, and so is dartos; the latter being a lower form of contractile membrane. Still further, the *elastic tissue of arteries*, where elasticity is in association with some low degree of contractility, places this structure in the category of contractile membranes.

In whatever tissue active vital functions, either of a chemico-vital, or cell-destructive power, are going on, there *mucous membrane* is recognized, purely and solely from its active vital functions, altogether irrespective of the form of differentiation it may assume; *i.e.* if in its totality it includes active vital processes that are not contractile processes, there the functions of mucous membrane exhibit a certain special active property; which declares what is its proper place in the grouping of the membranes in any special *tripartite membrane*.

On the other hand the pure *mechanical* and *physical* use of a tissue determines in all cases the metamorphic differentiation to be *serous*, whether that membrane be hard or soft, continuous or segmentary.

In the order of the animal economy, or in the extension of complexity, from low to high degrees of organization, the principle of the tripartite membrane is sustained, the number of membranes increasing as the organization is higher and more complex.

Thus the lowest forms, as Porifera, have but one tripartite membrane. 1. A silicon coat. 2. A very low sarcode or jelly. 3. An undetected tissue existing in the jelly, which gives the animal the power of alternately relaxing and contracting, whereby water, at short intervals, is propelled from pores extending from the internal mass to the surface of the soft body.

Man and mammalia are supposed to have ten or eleven distinct and special tripartite membranes.

1. The gastro-intestinal membrane (including serous, mucous and muscular membranes; and so of all the rest).
2. The broncho-pleural membrane.
3. The genito-urinary membrane.
4. The mammary membrane.
5. The vascular membrane.
6. The lacto-lymphatic membrane.
7. The ganglionic membrane.
8. The loco-motive membrane.

9. The integument.

10. The cerebro-spinal membrane.

This last membrane in many of its details is singular, and remarkably complicated; but it is identified in its highest function to be metamorphized muscle, and the adoption of this view was in MS. long before the same had been surmised by Fromman and Grandry, and also by Mitchell, of America.

Displacement of membrane is remarkably frequent, as well as differentiation of tissue.

Again, the special senses are recognised as seven, and each of these senses has a special tripartite membrane placed under, or subject to, its guidance and service. Thus the loco-motive membrane has distributed to it the sense of force or weight. The sense of touch has the integument placed at its disposal, and so on.

The *three* senses, smell, sight, and hearing, are viewed as wonderful mechanical expedients for inverting or abridging limbs, by which beautiful contrivances material is saved, whilst extension is greatly increased.

On the other hand, certain suggestions are thrown out in relation to the vegetable kingdom outside the field of true vegetable morphology, but only as it were to unite in one, the principles of mechanism and general laws, which govern both the animal and vegetable kingdoms, and moreover in some measure to suggest some points of practical utility in relation to the solidity of wood as indicated by the leaf.

The essential difference between the animal and vegetable kingdom, from the point of membranous morphology, may be briefly summed up as follows :—

Between the inner and outer wall of a vegetable cell, one or other, is much more active in its vital function than the one it opposes, and so gives an idea of serous and mucous membrane, *but not of contractile membrane*, nor of a nervous system of any kind. Hence a broad line of distinction is here given between the animal and vegetable kingdoms.

Much complaint will be made of the very brief manner in which animal morphology is considered. But if it were entered upon, somewhat fully, it would take up too much space, and sufficient is given for a simple outline or sketch. It comprises the substance of thirty years of careful observation and reflection.

EPIDEMICS.

It is laid down as a general state of the earth's surface, in relation to vital manifestations, that entire tracts or areas on its surface are subject to waves and patches of a limited and varied character which unfit it to sustain in integrity the essential conditions required for vital manifestation as a whole; and this is shown alike, in the vegetable and animal kingdoms, and these again, in their totality, tell upon man. For where vegetation is too rank, or meagre, or there is much dampness, want of pure air, and decay; fostering malaria—all these tend in themselves to create disease; whilst scarcity of food and clothing, on the other hand, tend to famine and depopulation.

Again, the evils of civilization, in the tenements used for aggregration, have a given depressing effect upon vital integrity or health, such, for instance, as relate to cleanliness, ventilation, and their collateral sequences in air respired, and food and drink appropriated, in an impure or adulterated condition.

Trades and vocations are not exempt from this general condemnation, in certain branches. Neither are habits or vices, which entail on offspring the residuum of their general morbific or depressing tendencies over vital power, any less exempt from the same censure.

Geographical position, and geological strata, have also a very marked effect upon localising defects in special organs, or constitutions.

These several agencies are in their nature more or less purely endemic, but the wider spread and all-pervading influence of recurring epidemics, is but the limiting of vital power in a consecutive and more general manner, by some form of disease which endemic speciality has already' endorsed, or which, through the law of change written upon Nature, it has failed to engraft upon any special point at its first appearance; but when once cast widespread by the winds of heaven, it never fails to appropriate some special localities as endemic haunts.

Of those agencies which tend to develop and extend disease, or promote health, as external agents to the earth's surface, the Sun takes the highest and foremost ground.

The Equator and the Arctic regions have the sun's rays alike in the entire year. The diurnal variation is extreme, and the *direction* of the sun's rays is very distinct between the Tropical and Arctic regions, though in the entire year the mere amount of rays from the sun is identical, upon any particular part of the earth, but the obliquity of the rays of the sun, or their vertical direction, is of the greatest importance in relation to vital manifestation.

Barometric pressure and humidity are equally important in relation to health and disease, but vital manifestation is in direct ratio to the vertical rays of the sun and the amount of moisture and barometric pressure. Hence, from a vito-physical point of view, these rank the highest in regard to the products of the earth, but not in regard to the health of man.

Though the spots on the sun have a certain relation to magnetic conditions on the earth's surface, as yet no impression upon vital phenomena has been traced as recurring with their appearance, or ceasing with their disappearance; therefore, as apparent agents in relation to vital manifestation, they may be accounted as *nil*.

Volcanoes and earthquakes, though frequent and violent, at times when epidemics have broken out, yet the occurrence of these terrible commotions within, which terminate in convulsions and changes on the surface of the earth, by no means runs parallel with epidemics. Whilst they are not viewed as causes of epidemic disease, but chiefly as coincidental, yet it is more than probable that they have an endemic influence attached to them, as immediately before and after their accession, dry, close, and sultry weather is observed for days, or much local electricity, and possibly emanations in the form of gases, &c.,

spread a morbific or depressing agency over man and beast *for a short period*.

Therefore, for persistent epidemic disease, such as plague, or cholera, volcanoes and earthquakes are of very secondary importance.

Upon the whole, temperature, moisture, and dead calms, have far more to do with endemic disease than any mere sudden eruption, or disturbance on the earth's surface; but for widespread epidemics something is wanted of a more general nature, slower, but more constant in its operation.

Leaving the conditions of epidemics, a few remarks may be made as to the peculiarities of epidemic disease.

The first and foremost of these is the power of *isolation*, which is so singular in some epidemics. Although according to an old proverb, " From the stall to the hall," is endorsed the fact of ailments in cattle, of an epidemic character, rarely existing for long, without spreading their baneful influence to man.

At times we have special diseases in poultry, and the rest of the farmyard healthy. Game, of the feathered tribe, in certain seasons will be diseased, vermin, as foxes, at another time, hares more rarely, yet never simultaneously. So, in blights and diseases affecting the vegetable kingdom, such as of wheat or any special kind of the cereals. In given years certain kinds of grubs or caterpillars will infest almost every apple ranging over entire tracts of country.

Disease of a special character has been in the potato for long; neither have onions nor yet turnips been exempt; and could man's knowledge scan far enough, he would find all bulbous and cereal growths, at times, subject to much disease, *but never several kinds at one period.*

In man, and the higher orders of animals, the isolation chiefly recognised is that of special organs. As in yellow fever, the chylopoietic organs are chiefly involved; in cholera, the chest organs and mucous membrane generally. In influenza, chiefly the lungs; occasionally a single organ is picked out with extreme exactness, as the spleen, in one part of Russia, in 1831. Since 1846 the chest organs in horned cattle have been specially attacked, and later on, rinderpest, has shown itself as a peculiar form of infectious diseases, namely, as an eruptive disease from blood-poisoning, but less special in its seat than any known epidemic disease, for scarcely an organ or surface of the body is free from its special form of elimination.

From these considerations it is assumed that vital force is always subject, in its manifestation as an epidemic, to some specific form of presenting its own power in any given tenement, and so to favour the law of isolation, rather than the withdrawal of equal degrees or amounts of force; and, by isolation, in a measure, suffering the machine to destroy itself by perversion of one or more functions, rather than by mutilating all consecutively.

In other words destruction to life, or limiting it for a

season, is accomplished by disease manifesting itself in special grooves, and over particular organs, rather than by any sudden and overwhelming catastrophe, as by freezing, or universal sphacelus.

This special form of evil gives both time and opportunity to man to do good to his fellow, and to be a helper against those evils with which he is beset, both remedially and by prevention.

The doctrine of isolation, as a law in epidemium, naturally leads to the history of a few special epidemic diseases of long standing, or of widespread diffusion. As for instance, from 1177 to 1817, plague spread to Mid-Europe and England, attaining its last hold in this kingdom in 1666; later on, at Dantzic, Marseilles in 1720, Vienna 1722, and Moscow 1772; since which time it has remained chiefly between Egypt and Asia Minor, and in these places since 1772 to 1817, to no very great extent, yet it shows itself occasionally for a month or two in one or other of its old haunts.

Small-pox, again, first made its appearance in England and Northern Europe about 1174 to 1177, and in America 1638, or thereabouts.

Leprosy, an old and venerable disease, being naturally a very chronic affection, and spreading slowly, did not show itself in England and Northern Europe before 1190, the same year as the third or Great Crusade was undertaken; but evidently, from the coincidence of time, it was too early for the assistance of infection by the men of that Crusade; and the seeds of infection, if that were required, arose from the previous Crusades of a very insubordinate position, from 1146 to 1187, to the famous one under Philip II., of France, and Richard I., of England, in 1190.

About the year 1600, or not far distant, leprosy ceased in our own country, and scarcely held its own in Northern Europe after that time; but in Spain it was well known in 1764, and later.

Hence, three well-known diseases, or four, as measles and small-pox, which were twin-brothers, born and fostered in Arabia, appeared as wide-spreading and infectious diseases not far from the year 1177 in England, and Mid or Northern Europe, at least north of the Apennines and Carpathian ranges.

These diseases each had for about 640 years, more or less, a prior distinct existence in the lands north and south of the *Mediterranean*, extending backwards from 1177 to 537. In 572 small-pox was not only known in Arabia, but had found its way into the literature of that country. Plague had broken out in the reign of Justinian in 543, or earlier, and again in 566, and had swept its thousands and tens of thousands of human beings from off the face of the earth; but there appears to be no authentic record of this disease reaching Mid or Northern Europe at these times; whilst leprosy showed itself in Italy in 614, and as a well-known disease in Spain in 714. How long, from its very chronic character, and, at this time, somewhat hereditary character, it must have existed before attracting general attention, it is impossible to tell; only it is quite sure it had lain dormant for some centuries, and became a common disease in Italy and Spain not far distant from 537, and onwards.

Plague, small-pox, and measles received a historical mention for the first time about 537, and, for want of further evidence, are viewed as having first obtained a true epidemic and infectious character subsequent to this period.

On the other hand, leprosy first came to Europe, or Italy and Spain, in the year 60 B.C., through the disbanding of Pompey's army at Brundusium, 61 B.C., which for three years had been overrunning Asia Minor. It evidently spread by infection at this time; but since 537 A.D., it has gradually become a less infectious, but a more distinctly hereditary disease, and now covering a wide-spread area from Norway to New Brunswick, the South Sea Islands, and Southern Asia, including India and China, and the West Indies.

The Mosaic Law gives rules to be observed by lepers, which are alone compatible with the supposition of its being at that time an infectious disease, and not in the slightest hereditary. Again, in the days of Elisha it was distinctly pronounced that it should be hereditary in the family of Gehazi; so that, from some occult circumstance, metamorphosis has been written upon this disease. In general, from Elisha's time till long after the time of Christ, it was an infectious disease, but strongly partaking of an endemic and hereditary character in its own proper focus, the borders of the Nile, from North to South Egypt. Since 1177 it has probably become less and less infectious, and more and more hereditary.

From a careful consideration of the three great epidemic diseases—plague, small-pox, and leprosy—an epidemic epoch or era has been assumed to recur about every 640 years from about 105 B.C. to 1817 A.D.

It is supposed that whilst any given epoch runs over a period not far short of 640 years, more or less, that it rather gains *in* extension up to about 200 years. At this point it remains moderately stationary for about 250 years, and then begins to gradually decline. The faint outlines which history supplies of the chronic disease leprosy, tend to confirm this general view. The same may be said of small-pox and plague, both of which are recorded in history a few years subsequent to 537 A.D.; but of this epidemic era we hear but very little till a short time before the next epidemic era of 1177. From this period the clouds of pestilence drop their baneful dews and showers over all Europe and Asia, increasing from century to century till about 1660, when the devastation they threw broadcast upon every city and country in Europe began to somewhat abate and, in relation to plague, to almost die out.

From 1348 to 1400 Black Death, *as a graft upon Plague*, added greatly to mortality in Asia, Africa, and Europe.

If such really was the case, then a new field of inquiry opens itself for further investigation.

1st. Can two diseases run parallel in the same body at the same time? Answer; Yes. As scarlatina, measles, and small pox: many cases of which have been recorded of late years.

2nd. Can two diseases amalgamate or coalesce, and out of that *coalescence* produce one new disease or *hybrid*. The answer is here given in the affirmative, and subsequent to this being written in MS; it has been advocated by Dr. James Ross, under the title of "The Graft Theory of Disease," adapting itself to Darwin's Hypothesis of Pangenesis.

Those diseases of an epidemic or an infectious character, which are here given as of a hybrid character, are the black death, syphilis, and the Athenian plague.

Black death is supposed to be the blending of an infectious lung disease, endemic in Tartary and the west of China, with the Levant or Justinian plague.

Syphilis and the Athenian plague are both supposed to be hybrids of leprosy under very differently modified conditions.

Syphilis is supposed to be the blending of plague and leprosy in one unifaction of disease, upon the whole, leprosy having the greater ascendancy, but nevertheless a hybrid between that and Levant plague.

On the other hand the Athenian plague is viewed as a hybrid of a certain rubeoloid disease, partaking in a modified form both of rubiola and variola, and this crossed with leprosy, the two poisons, as an epidemic, meeting each other on the confines of their natural borders, Arabia and the Nile, through the country of Ethiopia; and as a hybrid of germs, both arising out of hot climates, and of pure animal origin; like grafts upon particular stocks, under favourable conditions, are far more fruitful than either scions, when left to themselves. So when leprosy was grafted on a rubeoloid germ, the latter of which retained its individual stamp the most, an acute and wide-spread disease, of certain specialities peculiar to both, was the result.

In order to establish certain facts in relation to leprosy a general outline of chronology is given from the time of Abraham to the time of Christ. During part of this long period reference is made to man's age being 70 to 80 years of duration, from the time of the Egyptian Exodus, but longer antecedent to that time. Next, that the first general recorded epidemic which spread over the known world was about 767 B.C.

Again, that leprosy was always endemic in Egypt from the earliest historical records to this day, but it spread into Asia Minor several times before 767 or 750. That after 60 B.C. it spread not only in Asia Minor but also to Southern Europe, as Italy and Spain. But from 750, or thereabouts, until 103 B.C., leprosy outside Egypt was apparently a totally unknown disease, for which reasons are given why, if it had been outside Egypt, it ought to have been found among the Lacedemonians, who were in part of Jewish extraction. But in 800 B.C. it was in Palestine, and well known to the Jewish priesthood; yet shortly after this period we do not again hear of it in Asia, Greece, or in any place outside Egypt, until the time of Pompey the Great.

Attempting to go further back into the sources of disease and epidemics than the period history assigns for the founding of Rome, 753 B.C., some examination is made of the period of the Biblical record of the Noachian flood to the Egyptian exodus, this is called the epidemic of *human decay*, and as no idea is given of the diseases of men at this period, but only their *longevity*, it is further named the chronographic epidemic.

This epidemic lasted through an era of about 850 years, and is called the period of induction of human diseases. From the Egyptian exodus on to 745 B.C. is a period of about 746 years when diseases became established in such manner, as the decay of the constitution of man rarely sustained its vital integrity beyond 70 to 80 years. Subsequent to this time epidemic eras have been pretty regular in their recurrence, of about every 640 years more or less.

The most important of all subjects to establish is the veracity of the chronographic epidemium, inasmuch as it affected life the most markedly of all, and its proof lies upon the veracity of the Bible, than which no book is more doubted in its ancient historical details, and proof from outside itself is most desirable. To this end the great pyramid and relics of ancient astronomy are faintly glanced at, as affording corroborative evidence, and more to be relied upon than the doubtful products of pre-historic man.

Sir John Herschel is quoted as an author who was anticipated by the builders of the great pyramid in his conviction that records of science ought to be entombed in imperishable monuments, which goes far to show the advanced state of civilization in ancient Egypt.

Leaving for a while the history of epidemics we come to the etiology and the poison as a material agent, which constitutes the *materiæ corporis* of infection in epidemic diseases. Sir H. Holland appears to have first suggested an insect origin for cholera, and this was a great advance, because it widened our conceptions of active agents as living germs of disease.

Upon the whole, cholera is viewed as promoted in its development and spread by special kinds of *fungi*, whose pabulum is decaying animal matter, when outside the living body, and where decaying animal matter is found in a moist or wet condition, then, where cholera is present, we may expect its intensity to be very greatly increased.

These *fungi* probably act as a catalytic on the blood, and promote its decomposition, with depression of the heart and small arterial muscles, or partial paralysis. Agues, it is assumed, are produced in a measure by *fungi* also, but such as feed upon moist vegetable decaying matter, and especially *fungi*, which retain vitality in almost all seasons, but are most vigorous in damp, moist weather, and also from early morning dews.

An animal sarcode, or zooitic fungi, as they are here called, are supposed to be chiefly concerned in such infectious diseases as rubeola, variola, typhus, plague, spotted fevers, etc. Whilst for cancer and tubercle, the degradation of a higher class of tissue by some imperfect or depraved form of nutriment is the supposed efficient cause of these morbid growths or deposits.

A faint outline is referred to of the epidemic era in which we now live, dating from 1817 to 2457. First, as to the decline and fall of blood letting, which, started in 1823, grew less reckless and less, from 1833 to 1854, when as a rule, in Great Britain, it was universally condemned, and is now fast dying out in France, Italy, and Spain, etc.

The great cause of this change is considered to be the diminished force of the heart, as the chief organ isolated in our own epidemic era, and is that organ most under the sway of the metamorphosis of disease as implanted on our own special era; and because the heart, as a propelling power to blood throughout the body is feebler, therefore, the tendency to sthenic inflammation is less, and many surgical operations are much better borne on that account. In fact, the European constitution is much more approaching to that of Asia and Africa since 1850 than it was in 1800.

The result of this feebler action of the heart leads more to passive congestion in the capillaries, and serous, and saneous exudations, and less fibrinous.

The lungs giving, through auscultation, a better chance of observing changes going on in their structure, some reference is given to a certain lung affection with congestion, that has been observed with much attention in man since 1849.

Here the necessity of several long and forced inspirations is given as an efficient means of detecting pure congestion of the lungs from tubercle, pneumonia, or syphilitic infiltration.

A short outline is given of the same disease in an acute form in cattle; commonly called the pleuro-pneumonia of cattle, but really a blood disease of far wider range than the lungs and pleura; but since 1856 it has become much modified, or more restricted to the lungs and pleura. It made its appearance in Ireland not far from 1840, and about 1846 in our own land; proving *remarkably destructive*.

Yet it must be observed that there is a gradual change passing over the vito-physical resources, through the process of nutrition generally, which is slowly and imperceptibly altering, and in many cases, improving our vegetable and animal products.

Breeding and cultivation get the entire credit for the changes effected in our fruits, cereals, and floral gardening; and also in our horses, horned cattle, sheep, and poultry, etc.; but here a strong suspicion is thrown out that the change in the nutritive powers of animated creation have undergone a slight, but certain modified condition since the ingress of the present epidemic era, dating from 1817.

From this point a wider application is given to epidemic

epochs than that pertaining to improvements and changes in the culture of vegetable forms, or the breeding of animals, etc.; since their effects are traced in relation to dynasties, and the rise and fall of nations.

Acre, on the coast of Palestine, is taken as the centre from which in increasing circles, and wider spread areas from that centre, civilisation has been ever moving, and dynasties rising and falling from the time of Egypt's greatness to the present time.

The kingdoms of Egypt, Judah, Babylon, and Nineveh, Medo-Persian, Grecian, Carthaginian, Roman, Saracenic, and present European supremacy are taken under review, and from thence a close analogy is recognised between the spread of epidemics, and the length of their epochs, and that of the rise and fall of empires and dynasties, with many intervening peculiarities.

It is considered that the present anthropological era which began about 1817, taking Europe as its chief basis of extension, is spreading its leavening influence in a wider area from Acre than any previous anthropological era, and gives promise of knowledge and science spreading to every part of the world.

The human mind is taking a wider, deeper, and more general expanse than has hitherto existed, and implanted upon this is the universal adoption of the principles of *association and large combinations*, whether it be for commercial, scientific, or religious ends, or for pleasure, or aggressive purposes, as large combinations of skilled men in the art of war, large associations for political purposes, and societies, trades, and clubs without end, and all taking a different standard for their progress and stability to the societies of past ages. For in this era the press and publicity is the received standard, and secrecy is going fast into the shades of neglect and distrust.

The end of such association and free intercourse must be wide-spread civilisation, in which for all grades of mankind an equal code of civil rights must be accorded, or the end will be greater and more destructive wars than the world has ever seen. Moreover, the difference of races and colour has within it no sound ethnological basis for introducing class distinction, in according rights to them upon a moral basis, inferior to that pertaining to the white man.

The contrast of Empires, between the ancient and modern worlds, demand that equal rights be shown to men of all colours. The African and Asiatic nations and tribes are here briefly referred to.

A comparison is made between the rise, spread and decay of epidemic periods, and anthropological eras, with the view of vindicating the course taken of introducing anthropology in association with vital force, and showing that the moral force or power may, through its destructive creative genius, pervert and render useless to him the rich bounties which vital force gives to man, if security to property goes not hand in hand with judicious labour.

A more careful examination into the causes of epidemics is now given than at the commencement of the Sketch, from a reconsideration of the materials and data gone over, in which much recapitulation may be observed, yet for the end of viewing the same general facts from different points, this form of analysing is the most fit for a subject in itself naturally so difficult, and may be, in some measure, examined from a more general survey of leading facts ranging over a longer period of history than is usual in such matters.

Climate, heat, drought, mildew, locusts, heavy rains, earthquakes, volcanoes, trade winds, &c., are here briefly referred to; and in their relation to vital force, or its abstraction or perversion, as in the form of disease, these several agencies are admitted to have a very marked effect, but all of them more of an endemic, rather than of an epidemic nature. But such diseases as the Levant plague, or cholera, from their regular advance, and encircling the whole globe, and continuance from time to time over a long series of years, can scarcely be viewed as arising from changes wrought by an eclipse, comet, earthquake or volcanic eruption, or mildew, or locusts, etc., etc.

A brief examination is instituted of the change in temperature on the earth's surface from the time of Job on to our own time; or from 1500 or more B.C. to 1870; showing generally that heat has increased as we near the tropics and decreased towards the Arctic regions in regular gradations for 3300 years and more. That northern climates were warmer and southern climates cooler in the early history of the world, but during the progress of epidemic epochs they have undergone a slow but sure change.

Halley's theory of trade winds, &c., is *generally* admitted, but denying the sun to be sole regulator of them, because, if so, year by year they would return to the day and the hour, which is far from being the case.

An internal source of disturbance is suggested as a necessary accessory to the sun; to account for the checks to winds and monsoons on the one hand, and for the variations of cold and heat on the other hand, occurring daily, which, from being sudden, and not gradual, cannot have an origin in solar distribution, or in the action of the sun's rays on the earth's surface.

A short examination of Gilbert's and Halley's theory of magnetism as being a revolving solid mass towards the centre of the earth is given, and Dr. Barlow's and Haustein's of more recent times, and the general conclusion adopted is, that our chief disturbing agency to health, and promoter of disease has an internal rather than an external origin, viz.: that of an electro-magnetic nature, or, that within the earth, the electro-magnetic, or some like force, is disturbed, and subject to *consecutive changes* which affect the surface of the earth and its organized creation.

The secular variation of the compass from east to west is about 320 years, and back again another 320, so that a complete revolution is about 640, or the period here fixed, by observation from history, as being the period or epoch of an epidemic era. Some general reasons are given for supposing the historical data for inferring an epidemic era is 640, and the secular variations of the compass being also 640 years for a complete revolution, are not mere coincidences.

This inference is concluded by a general expression of belief that imponderable forces govern worlds and systems, hold in integrity the existing mechanism of the earth *internally*, and retain in integrity and perfect harmony the suns and planets beyond our earth, and the plants and animals living on its surface.

But as an organizing and creative power, they have as much to do *with starting or originating mechanism* as a fire has to do with constructing a locomotive engine, though the heat it supplies gives motor power to the mechanism.

This leads to the conclusion of the treatise by an examination of the epochal and detrital theory of geology, which is combated and doubted upon two independent bases. The first is, that the chemical constituents, as found in granite, are found in the sedimentary rocks, *in such excess and disproportion*, that the one cannot be derived from the other.

2ndly: That the mechanical effect of detritus must be to mix stratum with stratum, and their contained organic remains, especially those of small size, which are easily transported by water from place to place, but the opposite is the received geological theory, namely, that every stratum contains its own special fauna and flora, and when those of one epoch are found intermixed with the fauna and flora of another epoch or stratum, that, in such case, they are derived impurities, and the organic remains are NOT in the place they were first deposited as sedimentary rocks.

GEOLOGY.

There has been for long, say for fifty years or so, a sort of precise and yet reckless way of dealing with mundane antiquity. Thirty thousand and five hundred thousand years may have been allowed as sufficient time for the formation of some special stratum; and for another, upon some basis of analogy, or something equally general or equally precise, one million or five millions: a few odd millions of difference in duration in such mighty changes being unimportant matters, where time has to be calculated by billions or trillions, rather than by the million or the thousand.

The general tenor of such teaching has rested for the most part upon two hypotheses; the one being that of Detritus, from slow and successive waste from granite, with alternate stages of depression and elevation of this and super layers of rock, by forces sufficient to raise the earth bodily, if any one knew where to fix the fulcrum.

The other is based upon the fact that each stratum has *its own special kind of organic remains,* and which organic remains are never found in any other stratum; and if approximately found, they are of diminished or increased growth to that of some special stratum in which they are indigenous, or else modified in some particular manner, so that a good palæontologist could readily distinguish the aborigines from the strangers, or derived impurity.

The general lesson which was implied by such special aggregation of organic bodies, and also from the great lapse of time which existed between the end of one formation and the end of another, and so on, implied, for special organisms, a special preparedness to receive them, and before a sufficient state of preparedness could be obtained, adapted for great varieties of growth, long ages or epochs must elapse, till finally an age arrived which permitted all, or nearly all, varieties of form and species from lowest to highest in the scale of vital organism to live contemporaneously, though it is admitted in very different proportion as to size, number, and aggregate quantity.

Why one period of the world should be so uniform and specially adapted for a teeming variety both of plants and animals, and other periods or epochs appear to have been comparatively restricted to few varieties, seems, to say the least, wondrous strange.

Again, why the Creator had to fit up the world one way, and in another epoch has to fit up the same world in another way, appears equally strange, the more so, as for every epoch, adaptation was the prime point.

This implied progressive improvement in the adapter, only limited his skill and power; indeed, so patent has this appeared to some minds, that to them an adapter, or moulder, was not essential, and by leaving matter to itself, if only time enough were given to it, it would fall into proper mould and order of itself.

The writer, early in the year of 1864, wrote a pamphlet, entitled, "Doubts Relative to the Epochal and Detrital Theory of Geology," challenging the soundness of these theories, but discussing the subject in as few words as it was possible in such an extensive subject. Copies were sent privately to several distinguished geologists and others at the time.

It is reprinted at the end of this treatise for the purpose of suggesting that, vital force in its geological chart is anything but driven into a corner for evidence that its manifestations are necessarily gradational and progressive, any more than the organic bodies in the Laurentian rocks, or fossil man near Mentone, go to establish the same from a purely Palæontological point of view, or the recent researches of Sir W. Thomson on certain physical difficulties in relation to the same theory.

There are four points considered:

1st: From a practical point of view granite may be considered as the true base of all rocks, from the trituration and waste of which all others have been formed. Trap and basalt, etc., being too insignificant and, also, too recent in extrusion upon the surface of the earth, to be of any importance in accounting for the *materials* of either metamorphic or sedimentary rocks.

Both in the metamorphic and sedimentary rocks there is a marked discrepancy between them, in the elements they contain, and those possessed by granite. For instance, the proportion of lime in these rocks cannot be less than 15 per cent. of lime—may be it is not placed too high by making it even as high as 25 per cent.; whilst in granite the per-centage of lime cannot be higher than 8 per cent., but 5 per cent. will be much nearer the mark.

Again, carbon is not found in granite, or in the smallest amount imaginable, not as much as 1 in 1,000. Yet, in addition to the carbon, as a hydro-carbon, in our coal beds, and the carbonic acid in our atmosphere, there is at least four times more carbon contained in lime, in the form of carbonic acid, extending from metamorphic rock on to chalk, than is to be found in all other sources put together. These illustrations pertain to the positive side of the question.

But the negative side appears to be equally decisive upon the matter, for we find potash as much as 7 to 10 per cent. in granite, especially in felspar, yet, in the midst of such extensive disintegration, there is not so much as one single stratum of potash, or any salt of it, in any portion of the earth's surface, unless it be found in strata like to common salt. Hence it is clear that *granite* never was,

Geology.

nor is, the source of our existing strata, as the result of either slow or fast, or any other kind, of disintegration.

2nd: The small amount of *vitrified* rock in granite is opposed to the earth ever being a ball of fire, or a molten mass. Moreover, those forms of matter most difficult to fuse would be at the base, and the lighter portions, or more easily fusible, would be on the surface, of which we have no trace whatever, but much to the contrary.

On the other hand, the gradual rise of the land above the water, and land plants and animals appearing more and more as we near the tertiary strata or beds, show the expansive effect of heat within, *and not the contraction of the surface from the withdrawal of heat.*

3rd: The conformation of our mountain ranges and steppes, or successively elevated plains and prairies, demonstrate that the heat of the earth is not at its centre, but at a distance of about 800 to 1,000 miles from the surface, and in depth probably not above 5 to 25 miles, taking an average throughout.

From these considerations it is inferred that we never have had very much molten *débris* of any great depth issuing from the inner parts on to the surface of the earth.

4th: That organic remains, abide after their kind, each in their own stratum, prove that detritus never unsettled them; therefore, they constitute no basis from which to judge of age or duration. This fact is entirely conclusive against the formation of rocks from the *débris* or waste of previous rocks or strata, as the strata of the tertiary period cannot have been assisted by the *débris* of the secondary or sedimentary rocks, for they have no organic remains of the secondary strata, but only those which are peculiar to the tertiary, unless it be such organic remains as are classified by geologists as *derived impurities*.

UPON VITAL PHYSICS.

No writer has more correctly expressed the usually received opinions of physiologists, concerning vital forces, than Dr. J. Hughes Bennett, in his " Outlines of Physiology" (page 44):—

"In studying the different phenomena (whether physical or vital), physiologists are in the habit of using the term force much in the same manner as it is used by the general cultivators of science. Mechanics has its forces, such as that of the lever; Chemistry has its forces, like that of affinity; and Physical Sciences has its forces, like that of attraction. Physiology, also, has its forces. It has been supposed that, in the same manner, we have physical attractions and repulsions. Then we have contractile, nervous and generative forces. The idea of force, whether in physics or physiology, as explanatory of phenomena, must be regarded only as a theory, as a mental creation, which we employ as a convenient term to satisfy that intense desire of arriving at definite causes which is instinctive in man. On the other hand, it is often employed to express action which may be demonstrated and often measured. In this sense, it is as applicable to the action of the stomach, or of the liver, as it is to that of an electric telegraph, or a steam-engine." In short, vital force is often understood as a distinct and peculiar force of its own, directly antagonistic to ordinary physical force in many respects. Scarcely a greater dishonour could befall a man, than for him to suppose that vital force is but natural force placed by moulding or organizing special kinds of matter in a cumulative order of development, whereby each successive order of atoms appropriated to the primitive would add to the presiding force; and so, with increase of matter, comes increase of force.

This brings life down to gravitation, and places all in one series of uniform force, which at present cannot be demonstrated. First, let it be observed that in itself force implies *antagonism*, and as matter is viewed as an inert substance, variably affected by force, it would be as well to be explicit, since, for the want of plainness, persons differ who suppose that they entirely agree.

In the first place, gravitation as a law, as it is frequently called, and not the *attraction of gravitation*, is a misnomer in the sense of being a force—it is merely the result of a force, and in itself is no more an *active force* than a boiler is to a steam-carriage. The boiler confines and directs the force, but the active force itself is the *heat* applied to water contained in the boiler. All weight is in direct relation to *attraction*, which is the real active force. A pound of lead or cork, upon the hypothesis of every particle being equally attracted to the earth's centre, tells for certain that the amount of atoms is the same in both; but suppose a man weighed out twenty pounds of iron shot, and in the counter-poising scale his weights were made of brass, what a difference in the weight it would make, if under the counter, and directly opposite the scale containing the shot, there was a powerful magnet. It is easy to conceive, according to the power of the magnet, that the twenty pounds of shot might weigh twenty-two or four. If, instead of a counter, a large amount of magnetized iron lay, twenty to sixty feet below the surface of the earth, over a considerable extent of area, in such case the pendulum (if composed, in parts, of steel), would give a greater result than the exact amount of matter justified. But if it were composed in part of gold, and some large mass of matter of a diamagnetic character were placed in like manner, another source of error would come into play, though not to so great an extent as would arise from magnetized iron acting upon a steel pendulum.

The fact that greater density of matter over a given area is admitted, as affecting the pendulum in its oscillations, also leads to the examination of another source, not of error, but of misunderstanding, about which the most perfect accord is supposed to exist. Ask two men, perfectly conversant with the subject, "What do you understand by every particle of matter in the universe attracting every other particle, with a force varying inversely, as the square of their mutual distances, and directly as the mass of the attracting particles?"*

The answer given is one of the two following: Either that matter in one mass, which is the greater, attracts the matter in the smaller mass, not in relation to the amount of matter in the greater, but in the less mass of matter, because it is equal and mutual (in relation to distance), and the greater mass only attracts in proportion to the amount of atoms contained in the less, and no more, because, to repeat, it is equal and mutual.† Or the following: That it is inversely as the distance and directly as the mass; and that the smaller mass is, therefore, drawn towards the larger mass, not in proportion to the amount of matter which the smaller mass contains, but according to the plus of matter the larger mass represents over that of the smaller, and so the less

---

* Grant's "History of Physical Astronomy," page 26.
† See Airy, quoted at page 51.

is compelled by the greater in proportion to its mass. This latter is, no doubt, the correct interpretation of the facts observed in relation to the oscillations of the pendulum when tried at the base of large mountain masses.

But it still remains to be asked, Is the VIIth Proposition and its VIIth Theorem, in the third Book of the "Principia," correct according to the advanced state of experimental science or practical chemistry, which runs as follows:—
"Gravitatem in corpora universa eamque fieri proportionalem esse quantitati materiæ in singulis"?*

If towards the earth's centre is alone meant, then it may be safely granted that all particles of matter, when removed a given distance from each other, have, or may have, an equal attraction for each other, though this cannot be positively proved as yet. But the mean of the whole will certainly be in the centre, whichever view is correct; and, for astronomical purposes, the centre is the point in distance from which all objects ought to be calculated.

But when mutual attraction is insisted upon at any distance, then nothing can be more false than the doctrine of equal attraction; for the entire science of chemistry rests upon the *unequal attraction* which given elements have for each other, from the repellent condition in which electric currents, or caloric, places different material substances or atoms towards each other.

Boyle's experiment, so often cited, of a feather and a sovereign falling simultaneously to the bottom of an exhausted tube, connected with an air-pump, *appears* to settle the whole matter at once, and proves that every particle of matter is equally attracted towards the earth's centre. Such might do to settle men's minds pleased with toys. But remove all resistance to falling bodies, and the inherent

---

* Newton's "Principia," third edition, page 403.

velocity of the accelerating force is alone left, whether its degree of plus in attraction in one body, as compared to another, be as 1 to 1,000,000, or as 1 to 2. The amount of acceleration would be identical. But drop balls made exactly alike, but of different materials, from one height through a resisting medium, and first test them for porosity by the best constructed microscope, and the amount of acceleration between each would better test equal or unequal conditions of attractive force than each being placed in a non-resisting medium.

Occasionally the man of science indulges in the pleasing recognition of the consistency of scientific men in their modes of interpreting Nature, and recoils with well-measured irony against the man of literature, politics, or theology, contrasting most favourably his own adopted pursuits with those of the, said to be, less positive and more doubtful kinds of study. If the man of science were measured by the standard of consistency in relation to physical force, and votaries from the domain of two of the most exact sciences were canvassed upon this subject, the result would astonish some of the outsiders who came rather to learn than to criticise. Let one on either side be quoted.

"Our notions," says Professor Graham, "of the range of temperature acquire all their precision from the use of the thermometer. Cold, for instance, is allowed a substantial existence as well as heat, in popular language. What is cold? It is the absence of heat, as darkness is the absence of light."*

Contrast this with Mr. Grant's remark upon the tangental force. "The resistance offered by a body to move in a curvilinear orbit has been termed its centrifugal force; it is therefore equal, and opposite to, the resolved part of the

---

* Graham's " Elements of Chemistry," first edition, page 21.

centripetal force, which acts perpendicularly to the tangent. Hence, when a body revolves in a circular orbit by means of a force directed to the centre of the circle, the centripetal and centrifugal forces will be equal; but in every other case the latter of these forces will exceed the former, and will tend not to the centre of force but to the centre of the circle of curvature, corresponding to the infinitely small arc of the orbit in which the body is moving at the given instant. It is obvious that the centrifugal force has *no positive existence*. It merely arises from the resistance offered by the inertia of the body, in virtue of which the latter tends to persevere in a straight line."*

Now, *cold* is the condition in which attraction acts without the counterpoise of heat, and in chemistry it is nothing; *heat* in chemistry is the real force, and it repels particle from particle.

In astronomy, the repellent force, or centrifugal force, as it is called, is inertia—*i.e.*, do nothing, till a tap is given, or something starts off motion; and that tap is the *full extent* of all repulsive force from the first start until now, and though continually drawn from this tangent towards the centre, yet it is unsubdued, and is as fast as ever, trying to break off into the straight line of the tangent, from inertia or nothing at all; whilst attraction is the active force busily engaged in bringing about the ruin of this first tap, and though it has been at it for ages, yet it cannot, with all its activity, subdue this first slight tap or start. Hence the attractive power in astronomy is entirely master of the ceremonies, and the repellent power is nothing at all—it is a mere cypher, a nonentity; but in chemistry the repellent power is the lord paramount, and settles all difficulties under the name of caloric, and the attractive power becoming more and more developed as heat is with-

---

* Note in Grant's "Physical Astronomy," page 23.

drawn, is nothing. It will be said that cold is a term applied only to temperature, and, as heat or repulsive force is withdrawn, so cold becomes manifest, or is there. Quite so, but the attractive force is stronger as the repellent force diminishes. This is a perversion, it is said; attraction belongs to all particles, and is their state of natural inertia, but, when heat is given, motion begins, and the natural inertia is overcome by the force of repulsion, or caloric. Put it anyhow, the more attraction acts upon particles in the mass, the more are they in a state of inertia and nothingness. Next, in astronomy, the more the repulsion shows itself, either by its distance or velocity, the more it proves inertia; and the more the attractive force acts, as when three bodies are in one line, as sun, moon, and earth, so much the more evidence is there that a real force is in action, and not inertia in the slightest, for the inertia plays its cards all the other way,* and has a tendency to fall outwards from the line of the orbit, and that in a most determined manner, without the aid of any force saving the inexhaustible impetus in the line of its tangent, which was first given to it by a gentle tap—" Now go about your business." No one who has carefully examined the nature of the planetary motions can for one moment doubt the sufficiency of the explanation of motion by attraction proceeding from the sun as a centre, and that it thoroughly explains the *orbital motion* of the planets and satellites, if the preliminary proposition be granted, that the tangental force, once started, is incapable of being brought to composition and resolution by continuous deflections from the straight line, by the action of the attractive force.

That a planet, running in an orbit, obeys the order of increase

---

* It must be borne in mind, in speaking of cold and heat in relation to physics, that there is no reference to these conditions as applied to sensation.

and decrease of acceleration, as it nears or recedes from the perihelion, is simply a law of motion based upon the *direction* in which the planet goes, if its course is in an ellipsis; for if it were moving in a perfect circle increase of acceleration would not exist, but the radius vector (drawn from the sun to the planet) would not only sweep over equal areas in equal times, but equal areas would extend over equal distances in the line of the orbit in like times, and that without the slightest variation of time in traversing the line of the orbit throughout all parts of its course. This part of orbital motion is therefore freely admitted, and if not it does not matter, *for observation proves it*, whether admitted or denied; but the subject of inertia, as applied to tangental motion, is rejected on other grounds.

Inertia is complete passiveness; so that a motion started in a straight line continues in a straight line, *ad infinitum*, if unchecked or uncounterpoised, and this passiveness resides in all matter. Now, if motion in one direction is checked by that of an opposite, we certainly get the diagonal or resultant of the two, but we do not get inertia active to resist the continued central force *which gains by being neared;* and it is admitted that in one direction it has already given way to the diagonal, and in the diagonal direction it must again succumb in like manner from mere passiveness, and far from falling off from the circumference of the orbit it must be getting *nearer the centre*, according to the law of inertia.

Secondly, matter as inert cannot move in two different directions; it must go in one direction or another, not in two at one and the same time. Therefore, as the matter of this earth is at all times moving in a circle, by revolving round its own axis, every particle of matter in the globe is changing every minute in the direction of its tangent; therefore, it follows that it is impossible for the matter of the

PLATE I.

Page 45.

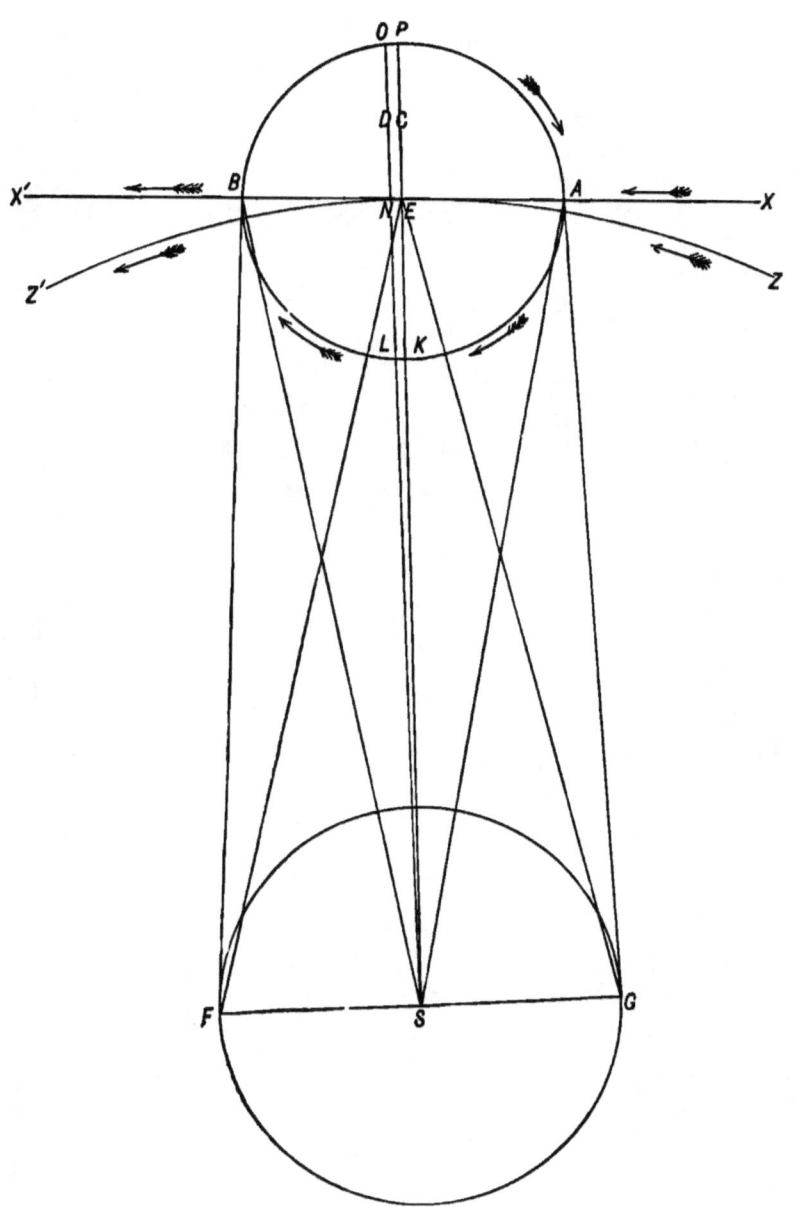

earth to be moving in a tangent, *always tending, to a greater or lesser degree, in a direction external to the earth's orbit.*

It will be said: This law of tangental force, or no force, is merely axoidal, and cannot interfere with the tangent of orbital motion. So far from its not affecting the subject, it is a subject upon which the entire doctrine of inertia rests, since, as we recede from the centre of a planet to its circumference, every particle of matter is increasing in its ratio of acceleration of motion, and, if not restrained by the force of attraction, would fly off at various degrees of acceleration in the line of the tangent the planet happens to be moving at the time, and in every direction, as dust from a carriage wheel when running with great rapidity.

Moreover the orbital motion and axoidal cannot both be west to east, if there were no force, but that of the tangent in the direction originally given, and from which it is by attraction constantly deflected, as may be easily shown by diagram No. I.

Let S stand for sun and E for earth, $Z Z'$ for the earth's orbit, and $X X'$ for the line of the tangent. Let A G and A S, E G, E F, B S, B F represent the entire direction of the attractive force between the earth and sun. Grant that the line X X is the line of direction of the tangent from X to A and then to B.

If, then, in all motion of the earth, in its orbit round the sun, or any other planet, the tangent takes the initiative, it is slightly in advance of the attractive force, or that the sun is attracting the earth *towards* its centre, but never directly into its centre, because of the ever-shifting position of the earth in its orbit, by reason of its tangental motion; then the centre of the earth, by such onward motion of the tangent, will be slightly in advance of the centre of attraction of the earth as given at the point E, whilst a straight line drawn from S to P through E would exactly bisect the earth into two equal parts, P C K A and P C K B, each to each. But the true axis of the earth by the motion of the tangent, shifts from the line X A E to the line X A N, and a straight line from the sun's centre through the earth from S to O bisects the earth into two unequal parts, whereof the section O D L B is less than the section O D L A, the amount of difference being included between the lines O D L and P C K.

As the attractive force acts equally through all parts of the earth from A to B when at rest, when in motion, the direction of attraction being from A to B, there is an excess of attraction from N to A plus that of N to E, by reason of the acceleration of the tangental motion from X to X', whereby in forces seeking their equilibrium the planet or earth E would be drawn from O towards A in the direction of A to K, and the axoidal motion would be in the reverse order to that of the line of the earth's orbit; for upon the hypothesis of tangental force being the initiative of orbital motion, and attraction the deflecting force which determines orbital direction, it follows that the axoidal motion of the earth would be in the reverse direction to that which is known to exist, and that it would move from east to west, which is not the case.

That it is viewed as the antagonism of an active to a passive force which determines the earth's orbit, and that that force is the tangental, a short quotation from Airy's "Ipswich Lectures" will suffice, page 80. Airy says: "The theory is this, that if we suppose the planets to be once set in motion (by some cause which we do not pretend to know), then the attraction of the sun accounts for the curved form of their orbits and for all their motions in those orbits." It is to be observed, that they are first started in the tangental course for them, or, after the tangental has started, to repeat then the attraction of the sun accounts for the curved form of their orbits. Again, page 81, *Ibid:* "The first law of motion is simply this, If a body be once set in motion, and if it have a certain velocity given to it, it will continue to move (if not acted upon by another force) in a straight line with unabated velocity. Hence, the law of motion by the tangent is the first or starting law of motion for the earth and planets in running along the course of their orbits."

By this very law of motion the axoidal motion would be in the reverse order of direction to that in which it is found. It is asked, But does not this go against the attractive force being the only force? Certainly, but in explaining the orbital motion the axoidal motion is never referred to, as though the two motions had no relation to each other, though effected by one and the same force. This absence of explanation is the more mysterious, as the axoidal motion is often referred to in works of astronomy, and is a matter universally examined by all practical observers.

Can anything be proposed as an off-set, supposing the tangental is only admitted as a qualifying agent in accounting for both orbital and axoidal motion? Here a suggestion, rather than a demonstration, is given without further apology. Let it be supposed that there are two forces of unequal acceleration; or, to put it in other words, that one obeys the law of intension in inverse ratio to the square of the distance, and the other, which is the weaker of the two, diminishes in the degree of acceleration *more slowly* than that of inverse ratio to the square of the distance, and *in proportion to the mass.*

This ratio of diminution requires a little explanation. Then let it be said that a force called attraction diminishes in intensity of force as it recedes from the focus of action in a fixed and constant manner, and also in a very rapid manner; but a force, here called the repellent force, diminishes in intensity as it recedes from the focus of action in a constant and regular manner, *but much less rapidly than does the attractive;* and as a result of this slower diminution in acceleration or intensity, it gets the start of the attractive, or is in advance in motion over the attractive, and always pushes the attracted body a little ahead of the curbing or attracting force.

To illustrate this, let it be supposed that S is the Sun and J Jupiter, or any other planet. Let G G G G represent the direction of the repellent force, and H H H of the attractive ; and let A B represent the line of equipoise of the two forces in relation to their nearness to the sun. Or that, at the line A B, the attractive and repellent forces have attained their perfect equilibrium, but inasmuch as the rate of acceleration is greater in the direction of G G G G than in the opposite direction of H H H, it follows that the increased acceleration will ever tend to project J towards the line C D, but the acceleration losing in force directly in relation to its distance from the sun, and attraction, by its slower acceleration but greater force, effectually checks the repellent force, when it reaches C D, and being no longer equally equipoised, brings J back to F E; for the ratio of acceleration in the adverse direction will, by inertia itself, be carried beyond the line of equipoise A B, but the increased resistance of the repellent force at E F will be too great for attraction to further overcome, and aided by the accumulated force at E F, rebounds back to C D; and in this manner, by two forces of unequal acceleration, a constant ingress towards and egress from the sun, are sustained in the form of perpetual motion. It is this unceasing oscillation, as here supposed, which tends to bring about that unresting state of the waves, present at all times, whether in storm or calm, as distinct from the tide, and is dependent upon

PLATE II.

Page 49.

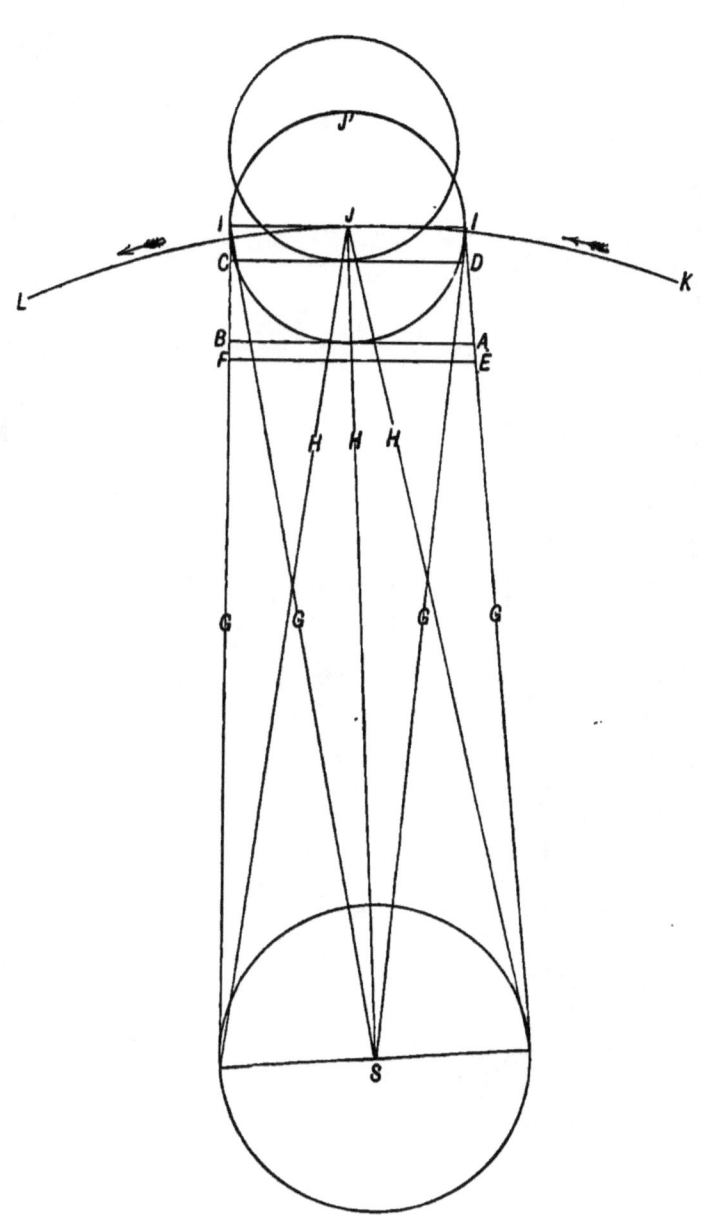

some condition external to the oceanic atmosphere. This, then, is the primary condition of motion in the planetary system, as the result of two forces possessed of unequal degrees of acceleration. It will be objected that if the oscillation is so great in proportion between the moon and the earth, or the sun and the planets, as from the line F E to the line C D, that our astronomical instruments would have long since detected it. Such, no doubt, would be the case, or even a 1,000,000th part of the implied distance. But the diagram (No. II.) is intentionally exaggerated for the purpose of more clearly indicating the kind of motion, and not its degree. With two antagonizing forces, as here maintained, let a tap be given to the sphere J at the point I, in the direction of J, and immediately the balance of equipoise is cast into a new direction, and in the onward motion, the sphere or planet, would ever be moving in advance of its true centre, and would describe a small arc in advance of the true centre, whilst revolving round its own axis and the sun. But inasmuch as these two forces perfectly equipoise each other, no new line of force will ever be lost, but sustained for ever according to the direction in which it is given, because the centre of equilibrium is not dependent upon the outside force or tangent, but upon mutual and unchanging antagonism *inter se*, and in whatever direction that antagonism starts in that direction it will continue, because there is no new outside force to interfere with the direction once started, whether that direction be from east to west or from west to east.

This view rests upon the assumption that axoidal and orbital motion are the result of the antagonism of real and active forces, and also that, without direct antagonism between forces, orbital and axoidal motion have no existence.

It need scarcely be said that outside astronomy it never was supposed to exist.

If it be granted that axoidal motion is obtained by two distinct forces, the repellent and attractive, naturally antagonizing each other, how, it is asked, is the orbital motion obtained? By a third motion or force, namely, the tangental. This force or *direction of motion*, it will be said, has already been disposed of as being destroyed by its direction being lost by the axoidal motion. Truly, such has been maintained when axoidal motion has to be sustained by one isolated force, the attractive, and the tangent left as the only remaining line of motion, which acted in antagonism to the centralizing direction of attraction; for in such case it has to fulfil the double function of axoidal and orbital motion, which motions are *in many respects* independent of each other, and perfectly distinct in their line of direction. But if the axoidal motion is already accomplished by the joint action of two distinct and antagonizing forces, the tangent only directing the antagonism in a fixed and constant course, which is the exact function for which it is called in, then all that is wanted receives its full accomplishment by admitting the tangental motion *to start the orbital*, and with it the axoidal, and no more.

If the two forces were of exactly equal acceleration, the tangental would simply produce orbital motion, for, no balance being lost by unequal acceleration, axoidal motion could have no existence; for axoidal motion for its production demands, not only opposing forces, but also that those forces be unequal in their degrees of acceleration, one moving more rapidly or else more slowly than the other.

That the repellent force is distinct from the attractive, in that it is in *proportion* to the mass, and not directly as the mass, one or two suggestions will be given tending to prove it.

But let the subject of attraction being directly as the mass, be first examined.

Airy thus speaks of it:—

"There is only one more point regarding the law of gravitation, on which I will here speak; it is the velocity or the change of motion which an attractive body produces on another body. I have spoken of attraction as if it were directed towards the sun, but we shall find that experiments of various kinds lead us to this conclusion—that every particle of matter attracts every other particle of matter, and that every planet attracts every other planet, that every planet attracts the sun, that the sun attracts the planets, and that the sun attracts the moon, and the moon attracts the sun, and that every body attracts every other body. Now the thing I wish you to understand is this: Suppose Venus and the sun are at equal distances from the earth, then the earth pulls the sun out of its way, just as much as it pulls Venus out of the way; the enormous difference of magnitude of the attracted bodies makes no difference in the movement which the action of the attracting body produces on them. If there are two bodies, a great one and a little one, and if something else attracts them, the great body is pulled through as many feet or miles in an hour as the little one."*

Perhaps there is no subject more difficult to comprehend than attraction being directly as the mass; the general impression is that attraction is directly *in proportion to the mass*, but to suppose the earth to pull the sun as much out of his way as the sun does the earth, at once dispels the illusion.

---

* Airy's "Lectures on Astronomy," delivered at Ipswich, 1848, fourth edition, page 106.

Let a different mode of illustration be given, that a clear view may be obtained as to what directly as the mass signifies.

If instead of using the words *directly as the mass*, it be enunciated that the *less of two unequal masses* represents the degree of attractive force in operation by the two bodies mutually attracting each other, then a better conception might be obtained; for the greater mass only attracts the lesser in proportion to the amount of matter the smaller of the two bodies contains, and the attractive force of the greater is only equal to the amount of atoms contained in the less, yet where the sun and moon and earth are in conjunction, as in the spring tides, their mutual attractions are greater than either alone; but this arises from there being two independent centres, both pulling and being pulled one way, combined with the relation of distance, and not from the direct increase of matter represented by the moon, added to that of the earth, simultaneously acting in one direction.

It need scarcely be added, that directly as the mass and inversely as the square of the distance only applies to large bodies, when removed beyond a certain and, as yet, unknown distance from each other; for the action of the pendulum is not only affected by large mountain masses *within a given radius* from the point of observation, but also by the density of masses beneath the earth's surface, whereby the motion of the pendulum is much affected, in areas no great distance from each other. A very important analysis of this subject has been given by Archdeacon Pratt, of Calcutta, on the deflection of the plumb line in India, caused by the attraction of the Himalaya mountains, and of the elevated regions beyond, and its modification by the compensating effect of a deficiency of matter below the mountain mass.

Vital Physics.

"On the Degree of Uncertainty which Local Attraction, if not allowed for, occasions in the Operations of Geodesy." By the Venerable J. H. Pratt, M.A., Archdeacon of Calcutta, 1858; and again, 1863, to the Royal Society.

In the repellent force, accumulation of force in relation to mass is admitted, or *that repulsion is in proportion to the mass;* and that, in relation to distance, it diminishes less rapidly than inversely to the square of the distance. To discuss this rigidly is not attempted in the slightest in this short chapter, but only the simplest illustration. If, then, heat represents the repellent force in one form, few would be disposed to doubt that its intensity increased in relation to the mass that is burning, *cæteris paribus;* but as applied to the planetary system in relation to the orbital and the diurnal motions, the following examples appear to bear upon the present subjects. From Brand's "Dictionary of Science," &c., 1852, Article "Planets":—

Planet.	Diameter.	Volume.	Diurnal Motion.		
			hrs.	m.	secs.
Mercury	0,398	0,063	24	5	28
Venus	975	0,927	23	21	7
The Earth	1,000	1,000	24	0	0
Mars	0,517	0,139	24	39	21
Jupiter	10,860	1,280,900	9	55	50
Saturn	9,982	995,000	10	29	17

The orbits of Mercury and Venus are less than that of the earth, each respectively, and therefore to be nearly 24 hours in rotation on the axis, or over that time, in so short an orbit, shows retardation in the axoidal motion in relation to the orbit. But in Mercury, in proportion to the orbit, the axoidal motion is much slower than in Venus, but the mass or volume is so much less; and therefore it suffers much from lack of accumulation of repellent power in itself towards the sun, and hence becomes slower in its axoidal motion. Venus is better, because it has greater volume. Mars falls short again, for as the distance increases, and the repellent power does not lose acceleration in equal ratio with the attractive for equal distances, the further from the sun with equal masses, and the greater ought to be the axoidal motion till the mean distance of our planetary system is passed; but the loss by volume accounts for this, whilst it explains the increase of axoidal motion in Jupiter, whose volume is so great, and the slight diminution in Saturn, with his decreasing volume though increasing distance.

After having so far discussed the subject of an *active repellent* and attractive force as bearing upon the planetary motions, the following propositions or assumptions will be given, with the object of placing the laws governing matter generally, and that of the planetary bodies specially, upon a more comprehensive basis, and in a form more capable of adaptation to the many conditions in which matter is found in the organic and inorganic worlds respectively.

1. That there are three entities, *two imponderable* and *one ponderable*, occupying and building up, as it were, all space.

2. That the two imponderables are fluids, which have equal extension throughout the universe, and interpenetrate each other, but are of *unequal acceleration*.

3. One fluid is the *attractive*, and is of constant and invariable acceleration, and is always inversely to the square of the distance, as far as is known.

4. That the other imponderable is *repulsion*, and of greater acceleration than that of attraction, but is variable; the acceleration is greater as it reaches the middle of our

planetary system, and slightly slower after it has reached the mean distance. (The planetary system reaching from Sun to Neptune.)

5. The repulsive fluid is also capable of abstraction and of accumulation within given, and, as yet, undetermined limits. This fluid is considered to be that fluid known in science as *caloric, electricity, and light.**

6. The third entity is *ponderable*, and exists in both a fluid and a solid state, occupying a part of space, and commonly known as *matter*. In itself it is variable, and resolves itself into elements and atoms, or small particles; the several elements or their atoms each possess distinct and peculiar properties.

7. As each atom occupies some portion of space, so it must possess *some form*, and as each element is peculiar and distinctive in its properties, so all the atoms pertaining to each element have a *special form*, each differing from the other *inter se*; and as a sequence to this it follows that each atom has an internal and external condition.

8. That the two imponderable fluids permeate all matter in two distinct ways:—1st. Every atom is permeated by one fluid chiefly internally, and by the other fluid chiefly externally. In one atom repulsion is plus internally and minus externally, and in another it is plus externally and minus internally, and so of attraction. But 2ndly, the

---

* Upon this interesting subject much new matter is given, and old well examined, by Dr. Tyndall. "Heat a Mode of Motion," 5th edition, 1875. Also see "The Radiometer," by Mr. Crooks; "Light a Mode of Motion."

degree of plus and minus in the atoms of the several distinct elements is dissimilar and special.

9. That the two imponderables diffused throughout all space are at perfect rest with each other, but being dissimilarly and unequally conditioned in the entity matter, there is ever a tendency to change and motion in matter, from the two imponderables ever tending towards equilibrium or rest. Hence, perpetual motion in the heavenly bodies, and constant change in matter, but especially manifested in animated nature.

What is here meant as inequality of permeation in degree or amount may be illustrated in the following manner. Take iron or steel, gold, strychnine, and hydrogen as so many distinct samples.

Iron is an element of remarkable tenacity and of singular variation in this respect, when subjected to different degrees of heat, but it is still more marked in the compound steel.

Let it be granted that iron possesses the attractive fluid largely, that it possesses it largely by central permeation of its atoms, and conversely, that the repellent fluid or entity is superficially applied, and in a small degree, the result will be two-fold. First, that heat will be slowly appropriated by each atom, whilst it is immersed in that entity or fluid in a very concentrated form, as in a very hot furnace, and therefore the approximation of each particle or atom will be slowly changed or it will be melted.

Carbon being an element in which the repellent and attractive fluids are about equally divided, and in which probably the repellent is the central fluid, different degrees of diffusion of this element in the iron gives to steel the special properties of brittleness and tenacity in a higher degree than iron possesses when uncombined.

The mode in which caloric is suddenly or slowly abstracted from the steel is the chief cause of the particles of iron being equally or unequally adjusted each to each, and therefore of the metal being elastic, or brittle and hard; as well as that it contracts unequally without air or gas, as it cools down.

Gold will illustrate the matter in another form; it is a soft, but remarkably ductile or malleable metal. How can this be explained?

Supposing the ultimate atoms of gold are superficially held together each to each by an external attractive fluid, and repel each other centrally, but that these two fluids are not equal in proportion each to each, but that the external attractive fluid or force is considerably plus that of the central repellent force, it follows that adhesion to each other's surfaces will be sustained to the level of their intact integrity, however finely beaten out; because the attractive force being plus on the surface, the edges of the atoms will adhere to each other, whilst the repellent being minus very considerably, under ordinary circumstances, the force applied from centre to centre of each atom will be unequal to destroy the plus of the superficial attractive binding force, and so fine continuous extension is sustained. Again, from the repellent force being small in each atom, it is slow to receive caloric in each particle, till that fluid is present in a concentrated form, as great heat, when it melts the gold. Hence the high degree at which gold melts, or its particles acquire central repellent force sufficient to overcome superficial attraction, and separate particle from particle. Yet, for strength, the attractive force not being central, toughness and hardness are not attained.

Put it in another form. Gold melts at a very high temperature, and is ductile and malleable; iron melts at a still higher temperature, but is very slightly either ductile or malleable. How are the two to be explained? First, gold is not strong or tough, but ductile and malleable, because its ultimate particles do not hold together from force at their centres; whilst, secondly, the superficial attractive force of gold is great as compared with the central force. Iron, in these respects, is the counterpart to gold.

Therefore, the particles hold each other from falling away from each other in gold with the greatest firmness and tenacity from superficial attraction, as distinct from central attraction. On the other hand, in gold there is no fulcrum of rest or adhesion, but, as it were, a very vacuum or want of individual bond each to each. The bond is that of any with any that may touch the superficies; hence the entire want of definiteness, or particularness in the direction of force, and this alone accounts for the universality of the extreme ductileness of the metal, and its indefiniteness in application to a specific or special end where strength is required.

The melting point between iron and gold rests in all probability upon the natural unimpressibility of each to the repellent fluid, whether superficially or centrally applied, both being sparingly impressed with that fluid, and of the two iron probably the least.

Let hydrogen be next examined. This metal or gas is, so far as at present known, the lightest. And why? We do not know; but, amongst other things, it is suggested that it is an element the atoms of which are (as compared with other elements) permeated with central attractive force, and superficially *very freely* supplied with or surrounded by repellent force, and that for this latter it has an intense susceptibility to its presence. Hence, it readily changes from solid compounds into the gaseous state, and each particle equally repels its neighbour from the rapid appropriation of caloric or the repellent force on its surface, whereby each particle is freely separated at equal distances from its neighbouring particles throughout the entire gaseous mass.

When hydrogen is gaseous, free caloric or positive heat is appropriated, so that the caloric between particle and particle is entirely appropriated to the one function of repulsion between particle and particle.

Hence, caloric is fixed and constant in each element, and each atom pertaining to such element, *cæteris paribus;* yet when not so conditioned the separate elements are, according to the conditions under which they are placed, proportionally susceptible to the influence of caloric, under one condition appropriating much caloric, and under another yielding up the amount of caloric previously held, as condensation either from pressure or chemical action; the chief disturbing agent to the attractive fluid being the variable proportion in which caloric can be entertained or appropriated by all kinds of atoms, whether appropriated centrally or superficially by any particular kind of atoms. The amount of appropriation of caloric is variable, according to changes in external conditions, but *constant* under *like conditions* in like atoms.

Again, a more difficult subject remains to be examined as illustrative of a type, namely, a compound radical, or, for the sake of precision, a "molecule," the word molecule being here used to denote a compound of affixed proportions. The compound radical or molecule, selected merely as by accident, falls upon the alkaloid strychnia—a vegetable molecule, the proportion of whose atoms is as follows:—

Carbon 21, Hydrogen 22, Nitrogen 2, Oxygen 2.

The several atoms of C, H, N, O, combined in a fixed

proportion as before indicated, constitute one organic molecule.

Suppose further, that each atom in this molecule be classified into central and superficial attractions, and carbon and hydrogen be viewed as having central attraction, and oxygen and nitrogen as being attractive superficially. Here, in the very midst of a new entity or molecule, degrees of force are acting towards each other, in response to each other's needs, so as to constitute a fresh centre of motion and action, but each elemental atom possesses the respective fluids in different degrees. Say, hydrogen is sparingly affected by the attractive force, but very susceptible to variations of the repellent force. Carbon possesses more of attractive force than hydrogen, and less of the repellent superficially. So oxygen possesses more of the repellent force centrally in proportion to the attractive force, and nitrogen is centrally repellent but superficially very feebly attractive (providing nitrogen itself is a pure element, and not a compound of silicon and hydrogen).

As some standard of acceleration for a fixed condition must exist, cæteris paribus, this standard is given to the attractive fluid, which fluid is considered to be invariable; and the same in every atom and molecule at all times *in the ratio of its acceleration,* the point of variation being taken from the repellent fluid, which can decrease or accumulate from any given point as a standard, as water at 60°, or oil at 60°. Caloric can be given up at this point down to zero, or accumulate up to 212° or higher, according to atmospheric pressure.

Hence the mutual attractions and repulsions constituting the inherent forces of a molecule at any given time may be entirely changed by variations in temperature, destroying all chemical affinities; the plus of caloric, beyond a given standard of heat, say 600°, being usually sufficient to destroy the mutual balance of antagonizing force and to obliterate the entity as a molecule, and to resolve it back to its atomic permanency, the caloric having outlawed the attractive element in the molecule by repulsion.

Having so far defined what is to be understood by a molecule, it will be well to next consider what is to be said of the molecule itself as an abiding element, or as a synatomic compound or molecule. It may be therefore said that each synatom or molecule is in itself to be considered as having distinct degrees of repellent and attractive fluids permeating its substance, each specific synatom, as of fibrine, albumen, protein, strychnine, and quinine having its own fixed and constant measure of appropriation of each fluid, according to external conditions, some being more attractive centrally and others superficially, but in the organic world so duly adjusted as to be pretty evenly balanced when in a state of perfection, but most of them very susceptible of the appropriation of caloric; the inorganic synatoms being frequently the reverse of this, as carbonate of iron, oxide of iron, oxides of metals generally, and earths, as carbonate of lime, sulphates and nitrates, etc., etc. But all synatoms, when solid, by absorbing caloric, are rendered liquid, and absorbing free caloric when fluid, become gaseous in one or more of their elements, or change directly to the gaseous condition.

It may be presumed that the properties of atoms are not only bound by the different degrees of contending forces existing in each particle, but by the special *form* each atom possesses. Thus, in crystallography special substances have their *particular forms* of crystals, as tartar emetic and arsenic, bichloride of mercury, silicon, carbon and iron, etc. If certain aggregate forms of matter have a specific range of crystal formation, does this arise from the absence or presence of form in the special elements or element entering into that particular aggregation? No doubt, *from form itself* being a special part of the constitution of every atom in each class of element, as of gold, silver, iron, lead, nitrogen, iodine, etc.

For how can it be that certain elements unite with other elements, as chlorine with lead, gold, copper, iron, etc., and in each of these its union tends to disintegrate, loosen, and, aided by water, to liquefy several elements of greater or less density and hardness, so that the inherent attractive powers are much enfeebled or rendered partially nugatory; but when added to a liquid salt of silver the mass separates itself from previous combinations, and falls down as a solid, the closeness of whose ultimate atoms, each for the other, excludes the water or the caloric from widely separating their respective atoms?

It is said that this is a matter of chemical affinity. Granted that it is a matter of affinity, is not iron capable of attracting carbon and oxygen with great facility, especially the latter, the same with sulphur and its oxides, and in each of these attractive force on either side binds the respective elements to each other in a close and compact manner, so that iron can closely attach itself to other elements, and so can silver, both as an oxide and as a nitrate? But iron refuses to unite with chlorine as a solid salt, but with silver, whose attraction of atoms for each other is by no means so strong as that observed in iron, unites to chlorine with intense affinity; not because it is an element of any strong attractive force in itself between its own particles, for, comparing silver with iron, it is not so tough as iron, and it melts at a lower temperature.

This contrast in the degree of affiance in two elements so opposed to each other can resolve itself only into a mutual affinity for each other, *powerfully aided* by the mutual adaptation of *form* to each other.

Why, again, are one or two elements so intermeddling with so many others as oxygen and chlorine, whilst others attract a neighbour to them with difficulty, as nitrogen directly with any of the ancient metals?

Not surely because it has no affinity for other elements; else what becomes of our cyanides and nitrates, etc.? Their presence speaks for itself, but if *form* aids in adaptation and close impaction, then the matter is clear enough, and no one can doubt this, if oxygen is carefully weighed with other elements. Its form adapts it for general intrusion, and with many very close impactions, and so attractive influence is aided by mere fitness, and such elements as nitrogen and chlorine have specific forms that, upon the whole, badly adapted themselves to other elements when brought into close contact.

Thus, if the primitive form of oxygen is that of a solid scalene triangle, it is but few forms with which it cannot edge itself in some way or other; but if chlorine be a solid equilateral triangle it is few forms with which it can well fit; but it would be much worse if nitrogen were an unequal ovoid as an egg, as for any real closeness of compact it would be almost, if not entirely, impossible—even a solid cube would be much more adaptable. Hence, in seeking to explain hardness, softness, and feebleness of attraction between elements, *form* appears to hold a certain and important position.*

---

* Since this was written, Professor Tyndall and many others have directed their attention to the ultimate forms of atoms, but as they do not in any wise change any leading point, the original description has been retained.

OF CHEMICAL AFFINITIES.

Form and unequal susceptibilities to two permeating fluids would appear to give all that is essential in determining special affinities chiefly known as chemicals. But here a source of great confusion arises—namely, how is matter usually conditioned.

There is the old attraction of aggregation, and chemical affinities; in other words, there are two forms of affinity or attraction, mutually opposed to each other.

First, of aggregation. This is an affinity of like for like, as the various constituent elements forming felspar or mica, dolomite, and carbonate of lime. None of these are simple elements, but they may be counted as synatoms or molecules of like elemental constituents, and as such they have mutual attractions which bind molecule to molecule.

So in the organic world the same thing exists in a very complex order, as the spiral tissue, pith, and true woody fibre of trees and plants each manifests an elective aggregation of like for like, just as in muscle the alternate structure is one form of like for like, and the external sheath is another form of the self-same order of attraction. Gold and iron, lead and silver, whilst pure and simple, have the same form of attraction set forth in its most defined form of like to like, or *similar attraction*.

This form of attraction is common to all great masses of matter, and metals after subjection to very high temperature; also to masses of matter which have been subjected to great pressure. In other words, it is a form of attraction that is the most abiding, and found in all great masses of solid substances scattered over the face of the earth.

So great is the tendency in Nature to *similarity of attraction*, that when large mounds of rubbish are made by successive additions and have stood for long, the several constituent parts are prone to arrange themselves in layers of special materials.

The late mound of the Vauxhall Gardens, when cut through some 60 or 70 years after its formation, showed this peculiarity very distinctly—clay, sand, pebbles, and small fragments of pottery ranging themselves in consecutive and singular order. Even a heap of stable sweepings, forming the midden, after two or three years, arranges itself into a manure of alternating layers of dark chocolate and green strata.

This form of attraction is that most commonly found in Nature, at least in the inorganic world, and widely contrasts with what is usually understood by chemical affinities, which will be here named *eclectic attraction* (ἐκλέγειν).

This eclectic attraction is very singular in its modes of manifestation; to bring it about perfectly it requires the component parts of which any combinations may be formed to be brought into a fluid state, either as solutions or gases; as tartaric acid and bicarbonate of soda, if kept comparatively dry, change very slowly indeed into the neutral salt Tartrate of soda, but dissolve each in water, and the action is almost instantaneous. But if a large mass, say one million of tons, were buried in the earth and subjected to great pressure, it is very doubtful if, after a hundred years, the salt would not be reduced to some more simple elements; just as the neutral salt sulphate of iron would probably revert to some sulphuret. Fluidity, and given ranges of temperature appear essential for the due manifestations of eclectic attraction. As living bodies are a compound of the *similarity attraction*, and the *eclectic*, it is deserving of notice that the primary part of conversion of raw material to be changed into animal tissue passes through a series of processes, as moistening, crushing, churning, and then suspension in a homogeneous fluid. All the crushed and churned matter in the stomach is reduced to a uniform liquid pap or mass, which in its turn is raised to a uniform standard of heat, and then it is passed on, to be presently taken up into the blood (and expelled in part as effete from the system) by a series of tubes and vessels — which, in the case of *veins*, let in much fluid and permit little to exude —and *lacteals*, which take up the more dense particles, each to remix in the double blood reservoir of admixture and purification—the heart and lungs—from which they depart to be diffused and to irrigate by self-appropriation all the tissues of the body.

Between entrance as raw material, and exit as effete matter, every particular particle has been reduced to successive changes in a moist or fluid condition, accompanied with a relative amount of fixed or free caloric ; and, according to the balance of materials and the caloric set free or appropriated by each particular particle, so are the changes in eclectic attraction ever undergoing successive orders of combinations and dissolutions till restored back to the inorganic kingdom.

That which takes place in organized beings in regular order of succession is, in principle, going on in the chemical laboratories of the manufactories of various kinds of material of human device and ingenuity.

In principle, the eclectic attraction is a series of atomic and synatomic or molecular changes, that, in their respective attractions for each other, show the greatest variety in their degrees of attraction, as in bone, membrane, muscle, and areolar tissue ; and which, with few exceptions, change their mutual affinities when placed beyond given ranges of temperature, or when placed under great pressure by surrounding substances.

No doubt the fluid caloric, in giving molecules and atoms

Chemical Affinities.

freer motion between each other, is one chief cause of eclectic attraction manifesting itself; also, by the elements appropriating certain proportions of this fluid, the relative bonds of attraction are disturbed, and dissolutions and recombinations of particles with particles is constantly recurring.

The views here given, *as the effects of two opposing fluids upon particles or atoms*, are also in harmony with the dynamics of the heavenly bodies, so far as regards their orbital motions, by their mutual attractions towards each other. Moreover they attempt to explain the effects of two fluids mutually antagonizing each other, and constituting distinct and opposing forces in relation to all matter *in each individual molecule or atom;* by which means a wide field is opened at once for the operations of forces, as applied to organized matter, as distinct from inanimate matter, and full scope is given for every form of matter and every degree of hardness, softness, and also for polarity of atoms and synatoms; neither do such assumptions interfere with the multifarious forms in which caloric, magnetism, and electricity—and perhaps light—mutually transform themselves into each other's places.

Whilst in organism it reserves to us the sun, as an ever-fresh fountain, to quicken on the surface of the planet the repellent force, by the action of his own force of repulsion upon the surface of our globe, in the form of heat and light, and by his spots apparently modifying our superficial magnetic currents; while the planets, in their acting again upon the sun in some form or other, restore back that force which is supplied to them by his influence, either in the form of magnetism or in the form of latent caloric, which acts and reacts upon matter without absolute loss of caloric to the sun, as is ordinarily supposed.

But no force, howsoever used, can explain the *mechanism* of organism. Force is but law—it cannot organize; and in all cases the being or object organized demands for its distinct form of manifestation the direct will of the Creator *in the first case*, no matter how forces act and react upon each other, after the organized matter has first received its stamp of reciprocated actions and reactions in the outward processes of development and decay from the seed to the seed again; for each seed has within it the stamp of the Creator's Will, and no germ is able to originate new germs distinct to those of its own primary conditions, the great apparent differences being the controlling effects of circumstances *in lower types or organizations; and that like genders like* is here embraced in its totality, with this remark, that after leaving the lower forms of animal life, the unisexual system in animals gives greater facilities for transmutation of species; but in lapse of generations, if procreation is retained, the progeny of dissimilar species reverts to one or other of the primitive forms, whichever condition it is living under, that is most favourable for retaining the type of one or the other primitive species in greatest perfection, and not to a new or distinct species; and so the persistency of species resolves itself back to that form of development in which it was first placed.

The first impress given to each organic form of life is, in animated creation, as much a stamp given to matter in a particular mould, as a likeness is to the die which impressed it. So an acorn is the first organic impress to the oak, and a pip to an orange-tree, and the spawn of a fish to its responsive salmon or trout. In astronomical science the primitive tap, or the duly-adjusted impetus given to matter at its start, from its own inertia, in the line of the tangent, determines by attraction the planets moving in their orbits, and so secures by that one tap the motion of the planets in their orbits for ever, and sets in motion the equally complex axoidal motion; and what the tap is to orbital and axoidal motion in astronomy, that is the original mould or stamp upon each series of molecular arrangement, in the germ known as perfect seed, to organic nature. In one system it is unifaction of motion; in the other the stamp of the germ leads to the unifaction of generation, but quite as distinct in one as the other.

It will be said that the opinion given at first was that it was one force, as vital force, which ruled both the organic and inorganic world. This is perfectly true.

Matter being subject to interpenetration by fluids having adverse action upon itself, makes the expression of those actions conveniently range themselves on the side of matter itself, as though the matter had in each of its atoms direct force or power, though the different *degrees* of susceptibility each element, in its atoms and primitive compounds, has for the respective fluids, are the direct and only measure of the force which it possesses; but as these are distinct and peculiar in each element, so the effect is uniform and invariable in every atom *similarly conditioned* throughout the universe, and that effect in its totality is the universal law of Precursion.

Or, that every atom of each distinct kind of element has a different degree of attractive and repellent power to every other atom with which it is surrounded or brought in contact, whereby all the grades of hardness, softness, elasticity, brittleness, opacity, etc., etc., are produced; and from this law of motion and rest all natural mechanism is sustained; though rest, abstractedly considered, is but a comparative thing, as absolute rest exists nowhere in this universe, for things apparently at rest are still moving, as the hand that writes and the paper inscribed are both moving at a great rate from diurnal axoidal motion, independent of other incidental motions. Again, the candle that burns has its matter in most contrasting motion. The stearine that is still an eighth of an inch below the flame is in gentle motion just at the line of junction with the wick, but when it reaches the lighted wick by capillary attraction, it is in much greater motion; but this is comparatively slow motion compared with the new compounds of carbon and hydrogen formed by heating the hydro-carbon, as well as much of the volatilized hydro-carbon decomposed. The caloric set free by these rapid changes in chemical compounds is, in a great measure, re-applied to increase the amount of space intervening between each atom of matter, and by such re-application is spent, in a measure, by securing greater repellent action, which ceases to be manifest at the point where the flame ceases.

The rapidity of motion, and the freedom, by increase of space, which each atom has whilst revolving on its own axis, especially if there be many different ultimate forms of atoms, will gender undulations of a minute character, and adapted for impressing the retina. Here the condition under which matter of a given composition is placed results in a uniform and fixed series of changes, and so would continue *ad infinitum* if similarly conditioned; but the ability to secure such results centres itself in the law of Precursion.

A tree brings forth its fruit—say, an apple-tree. The pip is beautifully entombed in its sarcocarp and epicarp; then comes the germ within the pip. This gorgeous encasing coffin, placed under the spacious vault of heaven, rots and liberates the pip, softened and swollen from moisture and heat; and the rain, or the tread of some bird or animal, aids in finding it a cover in Mother Earth. Here it bides the frost and snow of winter, and spring comes—then heat. Whence this heat? Is it caloric foreign to that gendered on this terrestrial globe, though called into action by operations going on in an object millions of miles off? No.

Doubtless the sun's calorific rays are identical in nature with that agent on earth, and liberated by chemical action, and known as positive heat or free caloric. By solar caloric the ostrich's egg is hatched, as is the pheasant's by sitting upon it, and so giving off animal heat.

This agent caloric starts motion in the pip—it is the repellent to attraction, and the matter in the pip or germ first begins to act and react from the decomposing matter of the pip itself, and, from the vigour gained by this appropriation, new and increasing actions and reactions on matter external to itself now take place; and these increased actions only cease when every reaction has completed itself according to the stamp given to its first germinal condition when entombed in its sarcocarp and epicarp, and every fibre, pith, liber, bark, leaves, and perfect circulation result out of the actions and reactions impressed upon the form of matter agglomerated together in the germinal spot contained in the pip, and first nourished by its own decaying matter, which is suitable pabulum for aiding actions and reactions in each upon the other from the first germinal agglomeration to the perfect fruit-bearing apple-tree, etc.

But it will be said, How are we to account for the doctrines of Morphology? It is by the action and reaction of repulsion and attraction, in the form of specific fluids of a purely interpenetrating character, that we are to account for the endless changes in matter adverted to in the manifestations of a tree's growth and decay, arising out of the stamp of the first germinal impression.

Or are we to say that the law of Precursion determines every pliant movement in the axial changes of a bud from the leaves to the petals, stamens, anthers, pistils, stigma, pollen dust, and granules on to the germ, carp, sarcocarp, and epicarp, etc., etc.; or, again, from the collar to the downward root and the upward stem, and to all the tissues comprising woody fibre, spiral and cellular tissue, etc., etc.?

This is a perversion of the meaning of law. Law retains order—it does not create; the law of Precursion sustains in their original integrity the order of successive changes, and that morphology which was stamped upon the molecular arrangements imprinted upon each particular kind of seed or germ by the Creator; so that, if a particular imprint were stamped upon a given germ, it would go on moving in that order of succession which would bring the individual parts and combinations, peculiar to the parent from which it sprang, back again to the germinal condition whence it started. The law of Precursion only tends to keep and preserve in a uniform and constant manner the form of growth imprinted first upon the germ from whence the parent sprang; and it is the office of this law to preserve this succession in the same plants, trees, shrubs, etc., etc., from year to year, and, as it were, to secure in constant action their genesis, first set light to by the expression of Will in the Creator.

Precursion does not retain the centres of ossification in distinct segments in the fishes of the sea, and so permit of a long season of progressive enlargement in all parts of the frame of the finny tribes of the deep; neither can the same law determine the early and rapid welding together or connation of the ossific centres of the feathered tenants of the air. These, and a thousand—yea, and thousands upon thousands of minute details resolve themselves into the archetypal plan of the primitive germ imprinted upon every organic product it has pleased the Creator to make.

So the planets in their orbits move in an ideal inclined plane, which is bisected at the centre of the extremes of orbital distances, and the centre of each semi-orbit is the point from which the planet may be said to start, as the centre of the perihelion, whence it runs down at an ever-increasing ratio of acceleration till it reaches the centre of the aphelion, from which point it moves within an ever-decreasing ratio of acceleration, till it reaches the same point in the perihelion from whence it started (and during the whole of the time the axoidal motion is not identical to the second in its duration, but is always suffering some small increase or decrease of time). The cause of this motion in an inclined plane is not very apparent. (Airy's "Lectures," pp. 83, 84.) But, probably owing to some peculiar internal construction of each individual planet as yet not determined, so in like manner the cycle of the year, unless artificially interfered with, is the usual time in which the circuit of morphological changes in plants undergoes every order of form, from the point of comparative rest and quiescence to the same condition again, saving in that order of growth in the vegetable kingdom known as cryptogams, whose order of growth and decay observe a fewer order of changes.

The more simple forms observe a much shorter period of vitality, but these appear, in many instances, to obey the influence of the moon rather than the sun, and are more of nocturnal and lunar periodic order than those of a higher and more complex organization. How far the lower forms of animal life, as madrepores, millepores, etc., or some of the porifera, etc., etc., are influenced by the moon does not appear very evident, but in the vegetable world lunar influence has been long observed, especially by water engineers, in the amount of vegetable growth in reservoirs as the moon gets towards the full. Though these phenomena are explicable upon principles of simple light and heat, yet they retain so close an analogy in the history of periodicity in some diseases, in relation to diurnal motion, as is manifested in neuralgic and nervous affections and fevers, that it is difficult to conceive the due adjustment of planetary motions to meet the successive requirements adapted to complete certain successive orders of change in a morphological point of view, and certain forms of disease,* without conceiving at the same time that one general law influences and binds together near and remote objects in one common focus of interdependency and correlation of actions.

Again, the sun gives us many rays and colours in one whole, and the earth, in organic nature, reflects those in distinct separate colours, above all, in a condensed form, in the Morphology of leaves; whilst that bright and burning luminary probably receives back in another form the rays he has sent forth, in some of the many metamorphic conditions which the antagonizing fluids assume in fulfilling their multifarious offices, as colour and heat transformed to magnetism and galvanism, etc. The effects are left behind in new combinations of matter; but the active agent is restored back to its original source, the sun.

---

* See Graves's "Clinical Medicine," page 436.

UPON LIGHT AND COLOUR IN RELATION TO VITAL FORCE.

From the humble attempt to acquire a more wide-spread and pliant order of force or forces, with their endless shades of affinities and repellents, than gravity can give, all blending into one great and general result—namely, the diurnal and annual motion of a planet round the sun, which is so adjusted that the light and dark, heat and cold, in their successive orders minister to the well-being of Nature on its surface—we next come into collision with the most delightful and beautiful of all subjects which can please the eye or fascinate the imagination. That subject is *light*, a subject interwoven with every shade and variety of vital force, as manifested in animal or vegetable life in all the shades and varieties of the never-ending series of colours, and their blendings.

Ethereal undulations and fine matter, or corpuscles, divide the field of honour in the court of theory in this field of enquiry.

One who cannot call himself so much as an amateur, and whose acquirements are so shallow, can scarcely be permitted to express any opinion whatever in such a matter. Therefore, leaving the subtle points of Biot, Arago, Hunt, Herschel, Brewster, and the marvellous tracts by Airy, etc., that which is here advanced is of the simplest character imaginable, and has relation only to *reflected* and not transmitted light.

First, whether correctly or incorrectly, the construction

of the eye, especially the human eye, has appeared to the writer to be more adapted, taking all points in review, to respond to undulations than corpuscles, or units of matter flying off from a large burning disc, such as the sun.

Secondly, whether corpuscles from the sun, as a finer polarized form of matter, or undulations, are the real cause of colour, both resolve themselves into matter in one way or the other; for whatever be the initial velocity of any particular kind of rays, *the angle upon which they pitch, and the fineness or smallness of that angle*, will have much to do in the way in which any particular ray is reflected or absorbed, which latter results in blackness; whilst other surfaces present a facing, whereby *every ray* is equally reflected, and not absorbed—hence whiteness, yet the form of ultimate atoms, either as elements, or when combined into given compounds, as gold and chloride of gold, iodine and iodide of potassium, give a *special* form in their simple or combined state, whereby given rays are well reflected, or rays cross each other and give complementary hues and colours, or they are imperfectly or partially reflected, or at such an angle that the reflection meets the eye partially, and hence dulness of lustre or hue, and feebly developed colour; so that whether special kinds of atoms come from the sun, or special atoms on the earth reflect on the surface rays of undulation from the sun, the result is nearly the same, the stationary matter or atoms upon the earth *produce the phenomenon of colour by reflection*.

Thirdly, mordants, having the power to fix colours, could not well do so, and keep them in the figures, etc., in which, as dyes, they are impressed upon cotton or muslin, if they only retained intact a particular *mechanical* form of surface, instead of maintaining a particular chemical compound, itself of particular atomic form and composition.

Fourthly, occasionally blind persons are found to have

remarkable powers of discerning colours by touch. As this can result neither from transmitted or reflected light, the special kind of surface is the only thing left for detecting the colour which presents itself to the touch.

It would therefore appear, in the case of the blind, that some fine impression reaches the touch from given colours, or surfaces possessing those colours, which nothing else imitates, and which cannot be supplied by the sun, so far as we can judge.

For if it be said that there is fine matter left on a given surface by the action of the sun's rays, in those things which do not transmit rays, but only reflect them, and from much exposure that matter from the sun has accumulated; on the other hand, it must be said length of exposure helps colours to *fade* considerably, and therefore material from the sun accumulating is scarcely feasible.

The view of a local surface of special ultimate forms from definite chemical compounds, forming molecules of precise angles and shapes, whereby certain rays are always reflected, and others partially, and some, as those at direct right angles, not at all, will give an insight into the reason why vital manifestation in every form of differentiation is accompanied by a change of colour in the organs and tissues of any given product, as the leaf, through its morphology, to the flower and the fruit; and every organ of the body and tissue, especially those of external manifestation, as feathers, hair, etc., are continually presenting to the eye endless varieties of colours and shades of hue, but in a pretty constant form of repetition in the wild state, from light, moisture, and food being, upon the whole, very similar; but under the care of man, these always suffer many changes by crossing with one man's stock and another; for no two men cultivate exactly alike, or feed alike, or protect from weather alike, or drain and manure alike, so that modifications in colour are constantly occurring in domesticated animals and farm stock, as well as in floral culture, from some one or other of these causes modifying the process of nutrition, and, therefore, of the condition of ultimate molecules, however slight that change, chemically considered, may be.

But molecule may refer to two, three, or more atoms constituting one primary compound or radical, which, as an organic substance, enters into the composition of one, two, or more animal or vegetable tissues or membranes; and each of such molecules, having its own special form, will reflect rays in its own special and distinctive manner.

Having sufficiently briefly referred to the subject of colour in relation to vital manifestation, a few remarks will be next given upon plasticity of cell architecture, arising from cell differentiation and morphology passing from primary cells on to membrane and tissue generally; or the changes which occur in the order of vital manifestation in animals.

UPON ANIMAL MORPHOLOGY AND DIFFERENTIATION.

As in floral hues and tissues generally, between one structure and another having separate colours, there is mostly some change in the chemical composition of the respective tissues, whose colours are distinct from one another; as the floral part is usually very distinct in its several parts from the leaf, and the petals from the pistil, and pollen dust from either; so, in all organic architecture wherever new forms of tissue or membrane require hardness, softness, elasticity, etc., to enable them to fulfil the functions to which they are destined, there we find, as the requirements may be, either the cell walls, or coutense, accumulating special material or definite chemical compounds to adapt it for the changed differentiation made, so that each cell may be fitted for its new destiny, as gelatine transformed to bone receives lime, etc., into its cell structure, and so a pliant cell becomes hard and resisting; but, moreover, cell differentiation, to adapt itself to cell architecture, undergoes every form of transformation which mechanism can devise to adapt itself to the circumstances required, so that out of the same primary impregnated ovum, in plants are formed pith, spiral fibre, woody fibre, parenchyma of fruit, and the hard structure of shell, as in the stone fruit, etc., etc.

In animals we have, from the same primary impregnated ovum, in the onward course of cell differentiation, serous and mucous membrane, brain and nerve structure generally, muscles, bone, integument, nails, hoofs, horns, and hair, etc.*

As cell differentiation and morphology in the vegetable kingdom have received such marvellous elucidation through Goëthe, by a series of careful comparisons, and subsequently from the patient researches of many microscopists and physiologists, of whom Schleiden may be said to take the lead, little more can be said than has already been said upon this interesting subject in its leading outlines, though in detail there is a large and ever increasing field of labour, unoccupied in various special branches, of intense interest.

But for the animal kingdom there has scarcely been any attempt to unravel this interesting but marvellous subject, and were it not for the probability of being able to condense, in a very brief manner, some general observations, which appear to have a very close relationship to the subject of morphology,

---

* In connection with cell differentiation, read the able article upon the "Structure and Growth of the Elementary Parts of Living Beings," by Dr. L. Beale, *British and Foreign Quarterly Review* for July, 1862.

as it exists in the animal kingdom, no outline would be attempted in such an intricate subject. But such as it is, in its leading outlines, is here briefly laid before the reader.

A general assumption is made that all cells are more or less formed of an inner and outer wall, here technically called *membrane;* and membrane in its lowest form is called *cell membrane.* This does not imply that the *inner wall* may not be continuous with and constitute a part of the soft or *granular part within* the cell, but in vegetables all cells and morphologies consist of *binary membrane.*

On the other hand, it is assumed that all cells of an animal origin have within them, or by differentiation have pertaining to them, the essential conditions of an inner, *middle*, and outer membrane, one of which is called *contractile membrane.*

In animals a plus membrane is added, or a plus material property is added to that pertaining to vegetables—namely, a contractile membrane.

The grand distinction, therefore, in cell morphology and differentiation, between animals and vegetables, is, that in material organism, vegetable membrane is essentially a *bipartite membrane*, and animal membrane is an essentially *tripartite membrane.*

What will best illustrate the nature of a tripartite membrane is the simple silicious porifera or sponges.

Granting the outer wall to be a more or less silicious deposit in the jelly-like substance of the sponge, and the inner contained sarcode, or jelly, to be the inner coat; yet by imbibition from without, as by capillary attraction, fluid is constantly and imperceptibly flowing from without inwards; but all through the pores and half tubular structure of the sponge, fluid may be seen thrown off at repeated intervals, from these open spaces filled with sarcode.

Here, in this low form of animal life, we see an action going on quite contrary to either syphonic action, capillary attraction, or exosmosis, etc. It is marked by intervals of intermission, and answers to nothing but a species of alternate relaxation and contraction; the effect is the interpreter of the cause. No direct observation can discover in the sarcode, or jelly, muscular structure, but the effect tells us that the fluid suffers compression at one time and is free from it at another time.

Again, in polypiphera many motions performed by these low types of animal life can only be explained upon the supposition of their having contractile power within themselves, though muscular structure cannot be detected in their organism.

Hence it is inferred that long before real muscle appears a contractile power exists—the result of material organism, which starts motion independent of the ordinary surrounding physical forces, though itself the result of special material adjustment of molecules with molecules of a given and definite character.

It is only after very great and highly-modified cell differentiation that we find cell architecture isolating the contractile membrane into a distinct structure, and assuming the true character of muscle.

Again, that which constitutes high and low forms of animal life appears to be, the *number* of tripartite membranes which are distinct the one from the other, in one living organism, and, by a wonderful adaptation in morphology, mutually blend with each other.

A brief enunciation of them, as belonging to man and mammalia, will be first given, with this proviso—that connective tissue, though the most useful in all the body, and ready to lend a helping hand in all difficulties as a supernumerary, yet in differentiation is at par, and is, in mammalia at least, the lowest form of tissue, and in the series of differentiations from primordial cells to brain tissues it is the lowest and most diffuse general tissue, which serves as general servant to all the membranes.

In true Mammalia it may be affirmed that they have *ten* membranes of a tripartite character.

Firstly: The abdominal viscera and glands generally from mouth to anus, consisting of serous, muscular, and mucous membrane, constitute but *one* tripartite membrane—the inflections of mucous membrane in the form of glands, as Brunner's, Peyer's, liver, pancreas, etc., being included as parts of the abdominal mucous membrane, and outside a second, or serous membrane, with unstriped muscle between as a third membrane.

Secondly: The broncho-plural membrane, with its minute and limited bronchial terminal muscular coat, but in some mammalia not purely constrictors or sphincters to the air cells, for they extend here and there to the larger bronchi.

Thirdly: The genito-urinary membrane.
Fourthly: The mammary system.
Fifthly: The lacteo-lymphatic system.
Sixthly: The musculo-osseous system.
Seventhly: The brain and arachnoid.
Eighthly: The veno-arterial system, or circulating system.
Ninthly: The ganglionic system.
Tenthly: The integument, which is viewed as a compound, or tripartite membrane.

These systems, or tripartite membranes, are most completely interlocked with each other, and also they are capable of the widest and most varied conditions of *transposition*.

The first membrane of a tripartite character, as here maintained, is the alimentary canal and its appendages.

From mouth to anus there is one continuous membrane, the mucous.

Its chief distinction is, the very varied forms in which it is found to infold itself as inflections, constituting the chief membrane, with connective tissue, of gland structure, as salivary glands, gastric follicles, intestinal glands, and the biliary and pancreatic glands. In all these there is a great proneness to cell *destruction* and *replacement* in their active functions.

Externally the mucous membrane has its own *serous membrane*, the peritoneum, which does not in all points oppose itself as a counter membrane, but is deficient in the buccal and œsophageal and rectal regions; but it follows it pretty closely in its abdominal connections, to carry out its *mechanical or passive function*, as a pliant smooth agent, in forming an extensive and ever changing *soft joint*. Here, then, the function is rather vital *mechanism* than chemico-vital or alimentative.

Between these two membranes is a third membrane, possessed of contractile motion—namely, *muscle*. At either end, by a law of displacement, where function requires it, the organic muscle gives way to striped muscle, as in the buccal, palatal, and pharyngeal regions, and in the rectal region likewise.

Secondly, the lungs are examples of the tripartite membrane.

The mucous membrane from the larynx, ramifying into all the smaller bronchi, ends in sacs, each of which has a sphincter, guarding its commencement, of organic muscular fibre—the sacs themselves, connected to each other, chiefly by connective tissues, serve as surfaces for blood vessels to ramify; and on the external part *the pleuræ* enclose the whole, and, like to the peritoneum, serve the purpose of a soft and pliant joint.

Here the proper muscular structure, which constitutes the contractile membrane, suffers great displacement, owing to the office it has to fulfil, in giving the slight initiatory start to respiration, by confining in the air cells the inspired air for one or two seconds, and then freeing it, which is afterwards carried on by the striped muscles of respiration belonging to the bony walls of the chest, or the respiratory muscles of Sir C. Bell.

By confining the air for an instant in the air sac, not only is the eliminative or destructive process aided by external motion ceasing, but, increase of heat expanding gas, when it relaxes again, gives an impetus to the exit of effete matter by elasticity.

Thus, under considerable displacement, we retain in the lung membrane the triple division of parts, and, as far as muscle is concerned, in a very limited and restricted degree.

It is not improbable that we may find the nasal mucous membrane, with the palate regions, a distinct membrane, with the levators of the palate as its special muscles; and that the rings of the trachea and larger bronchi are nothing else than very highly-differentiated and highly-modified serous membrane, fulfilling a purely mechanical function, and so far displaced and modified as to aid another membrane in the due fulfilment of its function; as the teeth are the millstones for the stomach, and the systemic arteries and veins are the food-distributors to the chylo-lymphatic system; the one being dependent upon the other for the due fulfilment of their functions.

The Genito-urinary organs are singular in their differentiations, and require in some measure a wider consideration than might at first appear, yet in itself simple and singular.

The oneness of mucous membrane which pertains to the

urethra, bladder, kidneys, the vagina, and the uterus need no discussion on the side of the female sex; neither in the male of the oneness of the mucous membrane of the urethra, bladder, ureters, and kidneys, nor yet of the vas deferens, and the tubuli uriniferi of the same sex, as those are matters of simple dissection and careful tracing. They are essentially continuous mucous membranes, even to the Fallopian tubes and fimbriæ of the female. The muscular membrane is found in the bladder and urethra, chiefly in the male sex, and in the uterus and vagina additionally in the female sex; but in neither is there any proper serous membrane.

For both the serous membranes of the ovaries in the female sex, and the tunica vaginalis propria and reflexa in the male, are properly considered as pertaining to the peritoneum of the abdominal viscera, and in nowise belonging to the genito-urinary system. But, however displaced and differentiated, it may be asked, Where do you find the serous membrane? Some would say, in the differentiated and homologous structures of the labia minora and clitoris of the female sex, and the corpora cavernosa and corpus spongiosum of the male sex; inasmuch as in one respect these may have a remote relation to serous membrane *in function*, since they are essentially *mechanical in their function*, and aid in accomplishing an end which is altogether impossible to be accomplished as an active vital function, unaided by a function which in itself is so entirely mechanical, and of independent and perfect adaptation; but these probably belong to the *Mammary system.*

Yet it must be borne in mind that neither male nor female sex are complete in themselves, and, by a peculiar differentiation, the ends of vital operation cease by their own independency of action; and not until the separate mechanisms are united by intercourse in the continuance of the successors to their own species is the perfect manifestation of the tripartite membrane completed.

But the ovum, impregnated by the granular matter of the male, in process of time is not only protected in an air-tight cavity, but soon a placenta is formed, and an investing *serous membrane*, though materially modified; holding within it fluid for the better protection of the growing fœtus, and also to render pressure to the mother more equable and less fixed. Here, then, we observe the ever-recurring fact, that the serous membrane is always devoted to a mechanical end, rather than that of an active vital function, as is shown in the development of the amnion.

True to its function as an excretory membrane, the mucous membrane of the uterus, after due course of time, in which the mucous and muscular membranes are undergoing a process of development and extension, in co-relation to a more ultimate growth contained within, which, by regular and equably sustained heat, aided by constant nourishment, arrives at perfection in about 273 to 280 days. A change now takes place, and the true excretory function of mucous membrane, aided by muscular force, begins to manifest itself. For a few days there is a greater or less amount of secretion or discharge, and a general lubrication, to be followed by most violent and effectual efforts at expulsion *per vaginam*.

It is probable that the exciting cause of this great effort arises primarily from a leaning in the mucous membrane of the uterine system to assume a special function, which is its legitimate function *when fully developed*.

This new function produces a secretion of a stimulating nature, and probably contains lime as one of its essential elements, beginning at its free or Fallopian end, and thence communicating stimulation to the ovaries.

It may be well to remember that in its analogue in oviparous animals and birds, the oviducts, by a process of secretion, obtain for the ovum the material for its outer shell; which, when completed, and imbibition can no longer go on, though the function is retained for the use of the following immature ovum, the active secreting surface refuses to retain *in situ* the living mass, which needs no further lime to form an external casement. The excess of that material, not being freely appropriated by imbibition, will act as a stimulant to bring into play nerve and muscular power.

That such is the case in the human being is inferred from a variety of incidents which have presented themselves in actual practice, but the enumeration and circumstances of which would trench too much upon space in so short an abstract as is the present; but one thing can be referred to without requiring much detail. It is the frequency with which calcareous matter is found infiltrating, sometimes slightly, and sometimes very much, into the substance of the placenta and the mucous surface of the uterus opposite to the placenta.

Occasionally it is so abundant as to convert the placenta into a semi-bony substance of considerable hardness and roughness, and, in detaching it from the uterus, requires very considerable nerve as well as manual dexterity. The great danger, of course, in such cases, is the violent streaming hæmorrhage.

Having said so much about the genito-urinary membrane, the next membrane to be considered is the Mammary membrane, which, in its transpositions, displacements, and differentiations, is as singular as any membrane hitherto considered.

The mammæ are singular organs, as bags and teats in cattle, swine, dogs, etc., etc. Though variable in their forms and modes of distribution, yet they are always in pairs or symmetrical.

The lactiferous tubes are inflections of mucous membrane, and the teats or nipples are supplied with a greater or less amount of erectile tissue.

The difficulty is to discover where its duplicate serous and muscular membranes lie; for, however displaced, the distinct membranes may be found, in some form or other, in some near or distinct locality.

Taking, then, the cervical fascia, which is but condensed connective tissue, it is found that this fascia spreads over and under the clavicle or subdermal ossified membrane, and thence it extends towards the mammary region. Also the same fascia dips down, and, running along the course of the large vessels proceeding from the heart, expands itself and envelopes the heart, where it constitutes the fibrous membrane of the pericardium, within which is enclosed the serous membrane of the pericardium.

This serous membrane is evidently in co-relation with the primary cervical fascia.

The heart itself, where it is double, is, correctly speaking, a symmetrical organ, but for obvious mechanical purposes is lodged in one cavity of the chest, (not necessarily the left); for if central the hard and projecting surfaces of the vertebræ would be decidedly objectionable to the freedom of its action, whilst its receiving fascia from either side the cervical region is suggestive of its equilateral origin.

As the fibrous coat of the pericardium by its origin proceeds from the cervical region, and is therefore an essentially external or subdermal membrane, so the contained serous membrane, with its accompanying fibrous membrane, must also be a subdermal membrane.

But in its symmetrical relations, as being the enclosed sac protected by the fibrous sheath proceeding from the cervical fascia, its primary seat or origin is from the fascia on either side the neck; and had the rudimentary ribs or true hæmal arches in the cervical region been fully developed, we might have had the mammæ in the neck and the serous membrane, enveloped by a strong fibrous sheath, in one or more detached positions, according to the site and position of the mammary distribution, and the heart altogether differently arrranged, but within the cervical hæmal arches.

But as it now is, the mammary glands, saving their being placed somewhere in the anterior aspect of the hæmal arches, have no fixed form of distribution. Ruminants have one locality, canine another, pachydermata a somewhat roving order of distribution, and man another. So, in the displacement of their serous membrane, though constancy is observed, yet it is far removed from the seat of its original place of distribution.

This singular membrane is essentially a subdermoid tripartite membrane, and carries with it a singular differentiation in its contractile membrane. This membrane, being outside the neural and hæmal arches, demands new functions, and partakes of a new form of differentiation in the form of striped muscle, and the *Platysma* is probably the simplest form of a true *subdermal muscle* in man, and is the true contractile membrane of the tripartite mammary membrane.

The mammary tripartite membrane is interesting upon another score than that which, as it were, introduces us to a morphology and differentiation in cell development most extensively used in the animal kingdom—namely, the striped muscular membrane.

For this membrane—namely, the mammary membrane—appears to be the last membrane added to the vertebrata, which introduces us into the highest class of animals; and its classification is based upon that membrane particularly, as the name indicates, its technical nomenclature being Mammalia.

Remove this membrane and we run down immediately to aves, reptilia, and pisces, or vertebrata with a thin covering of fibrous sheath over the heart, but entirely devoid of a proper serous coat.

It is, then, the adding of membrane to membrane in their ever widely differentiating forms, and the blending of each membrane's special functions, which constitutes the difference between the lowest porifera and the highest order of mammalia.

And as we remove one membrane and then another, so do we gradually descend from one grade in the animal kingdom to another. But it is probable, as in the insect tribes, that we have not, as the arrested conditions of high life, the permanent conditions of lower life in some almost unnecessary part; but that the presence of an entire membrane is presented to us in such an organ as the antenna of a butterfly or the sting of an hymenoptera, so that abortive or fragmentary representatives in lower life are ever presenting themselves as specific peculiarities in certain families and orders of lower life, especially in that of the insect kingdom.

We now turn to the more difficult tripartite membranes connected with the distribution of food for alimentation to the several tissues, and the destructive process connected with the removal of effete material.

The three membranes concerned are the ganglionic, lymphatic, and circulatory systems respectively.

We will begin with the Circulatory membrane.

It consists of a serous membrane, elastic tissue, and fibrous membrane, and beaded muscles lying internal to the fibrous structure.

The internal tubing of the veins and arteries consists of

serous membrane, with an external fibrous structure, which fibrous membrane appears to be the basement of the mucous membrane, and has suffered an apparent arrest of all further mucous structure till it arrives at the capillary system, to which it has been transposed. Here the peculiar active *cell-appropriating and eliminating* powers (which are supplemented by supplying capillary vessels to the muco-glandular structures in the great alimentary mucous membrane) indicate a close approximation in the capillary system to the function of mucous membrane, where much destructive cell change is ever going on, with an occasional economic ulterior end, as in the pancreatic and biliary secretions.

But in the capillary system the serous structure almost, if not entirely, disappears, and the capillary tubing is little else than enclosed walling, where active cell destruction is going on, and is abetted by exosmose and endosmose, aiding in the process of appropriating new and eliminating old material, between the moving blood current and the greater or less stationary structures through which it passes.

This, then, gives us a sample of partial membranous displacement in the capillary system, where the membrane, in its basement part, is running along in the veins and arteries, with its complimentary serous membrane, through most of its course, and the active functional tubing is placed half between the arteries and veins, under the title of the capillary system.

The *muscular membrane* in the circulatory system, both in arteries and veins, *occupies the middle coat*, but is sparingly distributed to them, unless it be at some particular point here and there; but the middle elastic fibrous coat of the larger arteries is probably composed, in a great measure, of differentiated contractile membrane in a low metamorphic form; for it is difficult to conceive, if some low form of contraction of a very abiding character did not exist in this structure, how it could remain in a contracted condition, and narrowing the calibre of an artery for long, and at another time leaving it dilated according to the amount of fluid passing through it, if the true beaded muscle had all the work to do, and in its action it was not supported by a slower and more abiding form of contraction.

The membrane just described gives an excellent illustration of the adaptation of tripartite membranes to other membranes of very widely different cell differentiation, since the capillaries reach almost every tissue of the body.

Added to the circulatory system is the Lacteo-lymphatic system, which is an economic and supplying system to the circulatory. Its structure is in many points similar to the former in its tubing, and ramifies almost every structure of the body.

Its internal coat is serous, its outer fibrous, and its middle is muscular, but sparingly developed.

Beyond its being a conductor of aliment through the thoracic duct to the circulatory system, by an entrance into the left subclavian vein near to its junction with the internal jugular, aided by the *ductus lymphaticus dexter* on the right side, it scarcely appears to perform any duty beyond that of imbibition of prepared aliment from the small bowels, and waste exudations from the capillaries of the system, or effete matter which is not returned into the general circulation, and which can be employed a second time to repair destructive processes.

But in the midst of this conveyance a glandular system, as mesenteric and lymphatic glands, intercepts the course of conveyance, and apparently subjects the chyle, or lymph, to some independent vito-chemical change, and may be adds, through a process of cell destruction, fresh material in such way and measure as produces in the moving fluid a gradual and nearer approach, both in colour and chemical compounds, to pure blood.

The chemico-vital function is the true function peculiar to mucous membranes, but in the complicated structure of these glands an outline of membrane is almost impossible; it rather appears to be a heterogeneous mixture of disconnected tissue, bound by fibrous and connective tissues; but in their function these glands partake of a decided chemico-vital rather than mechanical function, and are, therefore, viewed as transposed and differentiated mucous membrane, the outward fibrous membrane being the true basement membrane of lymphatic mucous membrane, and the glands but more active cell-destructive and chemico-vital organizers of the same membrane.*

The next membrane for consideration is the Ganglionic membrane.

This, then, introduces us to the most questionable and difficult of all we have as yet encountered, the more so on account of the peculiar notions of the writer, who is anything but orthodox, from a general physiological point of view, as to the real function of the sympathetic nerves.

Apart from morphology, or membranous differentiation, the special function of the sympathetic system has been viewed simply as a collector and distributor of *electricity*.

From whatever point it is viewed, its end is to produce *harmony* of action in the organic functions. Hence its indirect name of sympathy; and its grand plan of accomplishing that end is by economising waste electricity, and

---

* Some physiologists understand by basement membrane its active epithelial coating, or, in other words, its secretive or excretive function in the form of active and defined cell development; but when basement membrane is here used, it is a term to signify the sub-tissue of a more or less purely fibrous structure, over which is laid the superstructure of active cell development, often partially glandular, and always of a special cell differentiation.

expending it where it is required—at least, such is the notion here advocated.

With this preliminary explanation, an attempt will be made to describe its morphologies, and transpositions, and displacements.

It will be observed that the genito-urinary membrane, in its true morphology, is so far free from contiguity in its membranes, that it really has no serous membrane until the independent sexes are made one by conception.

Again, in the mammary membrane displacement is recognised in its fullest sense, and displacement is used to economise membrane in its completest form.

Moreover, it is doubtful whether the mammary membrane is not extended in the sexes to the erectile tissues, not only of the nipples, but also to the clitoris and the corpus spongiosum and cavernosum, as highly-modified serous membrane; and the mucous membrane of the glans as modified mucous membrane, ending at the commencement of the urethra, the muscular membrane being represented by the accelerator urinæ and erector penis; the excess of the mammary membrane in the teats, as an erectile tissue of the female, being compensated by the excess in the male organ of generation by the highly-modified serous membrane, as corpus spongiosum and cavernosum.

If, then, displacement of membrane in the several parts of a tripartite membrane is admitted in the genito-urinary and mammary membranes, it is equally applicable to the ganglionic tripartite membrane.

It is probably the fact that the lacteo-lymphatic membrane and the ganglionic tripartite membrane are complimentary membranes, and where one is absent the other is also, and they are both supplementary to the circulatory or the veno-arterial membrane, and are tributary and subservient to it.

To come to the long-deferred point, the ganglionic tripartite membrane consists of three parts:—

Firstly: The ganglia, as the cervical, prevertebral or thoracic, the solar plexus, and the greater and lesser splanchnic plexuses, etc., etc.

Secondly: The branches, or communicating fibres.

Thirdly: The muscular membrane, which membrane consists of the *heart* and the *organic muscular fibres terminating arteries*, or the capillary termination of arteries.

The ganglia are viewed as the glandular or mucous membrane perverted, yet in its differentiation retaining its function, but perverted in its physical character *in toto*. The communicating fibres are serous membrane in function; and the heart and capillary muscular fibres are the true contractile membrane belonging to the ganglionic system.

The enormity of this morphological change and displacement will be greatly questioned; probably it will be the point at which many will say we can no longer go with such extravagances.

But, strange as this explanation may appear, it is more than probable that it is abstractedly correct. For if we take into consideration that the lower forms of the subvertebrata have no real heart, neither have they any lacteo-lymphatic system, nor yet ganglionic system, the feeble circulation and the power through imbibition and exosmosis, especially where the circulation is feeble and slow, need no complicated system of elaborate digestion or conversion of aliment into blood, nor yet any complex and perfectly rhythmical and systematic form of carrying on the distribution of aliment; or a beautiful system for removing and economising of old and used-up material; but in higher life these complex systems are essential to the well-being of the individual, and cannot be well dispensed with.

The first question that arises is, Why should the heart

have ramifying through it such a progressive increase of ganglia, and nerves distributed from them?

In a system, as the human body, we have thorough domestic economy throughout—to wit, where old tissues are re-used, as in the lymphatic system, as well as in the biliary and pancreatic; yet, in relation to *electricity*, what provision has Nature employed to turn an active agent into genuine utility, such as free electricity, since for waste exuded material a lymphatic system is in full operation, and is of the greatest service? But what is done to get rid of superfluous electricity set free by chemical action?

The answer is that a system is at hand and in constant operation for this very thing — namely, the ganglionic system; and its function is to collect electricity, and at its ganglia to transmit it, after modifying or changing its course, so as to direct every slight change in degree, intensity, or condition of electricity, into its right and suitable channel from ganglia to ganglia, until it has met with its final distribution at the ramifying ganglia of the heart, and also the minute muscles placed round the termination of small arteries.

But from all we know, mechanical pressure is always sufficient to excite muscular contraction in organic muscular fibre, but two other agents also greatly influence it.

1st: Electricity intensifies muscular irritability, and so greatly aids *regularity* and *quickness* of action.

2nd: Temperature greatly aids it; when above a given point the action of the heart is increased, and below a given point it is greatly depressed.

The two great points of distribution, then, appear to be the heart and the muscles surrounding arterial capillaries, and, being under the guidance of one system, they will always act in sympathy with each other, and in health in harmony with each other.

Concerning their distribution to the capillary termini of arteries, nothing is known with certainty; but with regard to the heart, the distribution of its ganglia and nerves indicates an arrangement of a most comprehensive character for complete diffusion throughout that organ. For our more precise knowledge about the nerves of the heart, we are chiefly indebted to Dr. R. Lee and Dr. James Pettigrew.

It will be perceived that the endocardium, which is continuous with the arteries and veins, has no relation in its muscular membrane to the heart, but the muscular coat of the heart is the excessive development of a distinct tripartite membrane — namely, the *Ganglionic muscular membrane;* and this membrane is, as it were, wedged in between the junction of the spent and depurated blood and the new blood by the lymphatic system, and the fresh blood from the lungs made ready to go its round of vital operations between waste material and fresh matter to the several tissues of the body.

If it be true that ganglia are distributed to the arterial muscular fibres at their distal termini and at the heart, and that this membrane is really an economiser and distributor of *electricity*, then its freedom of action and independent origin as a distinct tripartite membrane can be easily explained.

For, in such case, the heart will be ruled in the degree of its action in sympathy with the body generally—not from the *kind of blood* flowing through it, unless when the kind also extends to the quantity; but it will be more influenced by the effect which blood has in relation to chemico-vital, and therefore electric changes which are going on in remote parts of the body, and thus tone down or rouse the heart up to more energy, according as these changes modify the electric equilibrium in morbid function and nutrition; and so local electric changes will become early notified to the heart before the mass of blood suffers any important change. Moreover, the minute arteries will be depressed or excited in their muscular walls by the local condition of electricity, and the part itself will be the first partaker of any change of circulation of a retarded or accelerated nature.

This would well explain Haller's observations on the increased local circulation in inflammation; for if it began to increase locally the current of blood in the arteries of the affected part or limb by stimulation and dilatation of them, then, as a sequence, the venous blood would become proportionally increased in amount, and end in *local* plethora.

But if the heart and vessels were all one membrane, the more active function of the heart must be more or less in abeyance to, or in harmony with, the slow contractile powers in the larger vessels; and the contraction and dilatation of the arterial system must keep pace with the quick or slow action of the heart, which would subject the circulation to too sudden and extensive changes, and would prove a source of great danger.

We now leave the membranes more especially concerned in vegetative life and turn to the tripartite membranes of animal life.

The mammary tripartite membrane has already been examined and viewed as *essentially transitionary;* whilst the veno-arterial membrane was considered an essentially diffused vegetative membrane, and fulfilling the ordinary chemico-vital function of a mucous membrane in the ramifying capillaries pertaining to muscle, bone, and nerve belonging to animal life, and other membranes—in fact, to all membranes external to itself; and when, from peculiar circumstances, the ordinary functions of chemico-vital forces are inadequate to minister complete efficiency whilst supplying the wants and removing the *débris* of the essential organs necessary to sustain a complete equipoise between waste and supply; then the veno-arterial membrane plants reserved stations to act upon emergencies, as a species of temporary scaffolding, or refreshment stalls, which are by no means essential to the well-being of the system, in the form of the vascular glands of Paget, in which vito-chemical changes, and probably certain mechanical advantages, as diverticula, are secured, whereby sudden variations in function, or more chronic conditions of disease, or disturbed function, can be better sustained.

But we now leave, in a great measure, chemico-vital changes, and those conditions which are essential to secure vital integrity, to consider those higher and more ulterior ends which bring vital force into direct antagonism with the physical conditions by which it is surrounded, by opposing counter-physical and mechanical agencies *versus* gravitation.

We begin with the *somatic*, or body senses—more especially the senses of touch, *and force or weight*, or the muscular sense of Sir C. Bell, and the sense of want, or hunger, which latter has elsewhere been called the *hæmal sense*, and is so closely connected with the ganglionic system in many parts of its distribution as the par vagum, or pneumogastric nerve, which sensory nerve is the first of the somatic senses in the order of appearance in the subvertebrate kingdom.*

Each sense has superadded to it, its own proper tripartite membrane or sense apparatus, which, for completeness and distinctness, stands in bold contrast to all previous membranous mechanism and cell differentiation; but in its first metamorphic change it approximates in some measure the system of metamorphosis from which it has emerged.

---

* "An Experimental Inquiry into the Existence of a Sixth Sense," by R. F. Battye. *Vide Edinburgh Monthly Journal of Medicine* for February, March, and April, 1855, and *Edinburgh Medical Journal* for February, March, and May, 1859.

The *par vagum*, or *hæmal sense*, is supposed to give consciousness to its owner both of *hunger* and *suffocation*, etc., according to the state of the blood in that particular organ to which it is distributed, in contrast to those senses which give consciousness to pressure, resistance, and external stimuli. Hence, in determining the function of this sense by irritants and mechanical injuries, only one side of the question is settled, and that is the negative one, or the want of action; nay, even the par vagum may be cut, and the cut end of the nerve on the proximal side to the brain may give sensation to the central mass as though it were still intact, and acts may follow as if it were in its natural condition, so far as eating and breathing are concerned. Just as men with an amputated leg often speak of the toes itching or the limb paining, which no longer exists as an integral part of the body, so after section of the par vagum acts follow from mere sight, which are dictated by central sensation. Certain states of blood, as of given conditions of acid—as of carbonic, lactic, hydrochloric, etc., etc., which has not yet been fully excreted from the capillary network—are supposed to be the natural stimulants to the par vagum in exciting the feeling of hunger and suffocation, etc., the one pertaining to the lungs, the other to the stomach. No doubt the heart, liver, and small bowels yield certain morbid feelings when their functions are impaired from some altered condition in the capillary system of the organ affected.

The par vagum, then, is considered to be a true sense, placed in proximity and juxtaposition with the other senses in the great central mass—the brain.

This sense gives information to the brain of certain wants and states belonging to the organic, or vegetative conditions of the body, more especially that of the stomach and lungs, but leaves to other senses information beyond this part of bodily framework.

The senses of touch, and of force, or weight, give information of the surfaces of bodies, and also, in the sense of force, their *degree of resistance and proportions of weight.*

The weight of bodies and degrees of resistance to muscular force are measured, as is here supposed, by one sense, commonly known as Sir C. Bell's muscular sense, but it has been also maintained to reside in and about *joints;* but not solely here, as it appears to have a partial distribution on particular parts of the integument, as at the bend of joints, the palms of the hand, and the soles of the feet—in fact, to all those parts of the integument which, when wetted, feel the cold from evaporation or cold liquids the most. The sense of touch is more adapted to recognise the roughness or smoothness of bodies and their equable warmth, both senses embracing, as part of their function, states of temperature. The sense of diffused touch recognises heat the most quickly, and the sense of force, when distributed to the integument, recognises cold the most quickly; in this country between 56 to 80 degrees Fahrenheit being about neutral ground, but above this point the back of the hand quickly recognises increasing heat in water, and below this point— that is, below 56 degrees—the palm of the hand most easily detects falling temperature.

Mention scarcely could be made, in so brief a sketch, of such matters as heat and cold, or even of pain, as having special senses for each; but as there is a growing tendency to increase the number of our senses among physiologists, and those of the highest cultivation in this branch of medical knowledge, a reference to it appears necessary, the more so because a careful examination of details leads to a supposition that special senses carry with them certain collateral elements, that appear to be possessed in different degrees by *all* the senses; as, for instance, that of pain and also of heat or cold, each having a certain higher, or lower, perception of one or the other, as their particular cerebral impressions and anastomosis with each other in the great central sensient mass may impose upon each individual sense.*

Certain senses, as the optic and acoustic, appear to be specially free from any marked changes of temperature, saving during fevers, etc., when their functions, with that of the cerebral mass, frequently suffer considerable change. Also from indigestion, and the presence of certain acids and other compounds in the stomach and bowels, unpleasant sensations of heat are experienced, the result of impressions on the periphery of the par vagum, etc.

OWEN considered that the special senses carried with them certain anatomical peculiarities which entitled them to the nomenclature of *sense capsules.*

It is almost a pity to alter nomenclature to fit it to any special theory, or new explanation; but inasmuch as *capsule* is too circumscribed in its application for the present subject, the nomenclature of *sense apparatus* though rather indefinite in its signification, will be used in preference to sense capsules, for it is sufficiently expansive to admit any of the senses into its category of objects.

Leaving, then, the precise object for which, in the differentiations of the notochord, the nomenclature of the arches is adopted, as that of the hæmal and neural arches, it is here assumed that these arches are the apparatuses of three senses—the senses of force, touch, and want, the two former of which inosculate and intersperse with each other at several points, though at no point do they amalgamate so as to constitute one sense; but they frequently occupy the same ground, or area of surface, by the interpenetration of their fibres. So,

---

* *Edinburgh Medical Journal,* 1859, p. 797.

in diffusion of their fibres from their spinal roots, they occupy a common neural arch, or the posterior arch of the notochord.

On the other hand, the sense of want, elsewhere called the *hæmal sense*, and giving consciousness both of suffocation and of hunger, occupies the anterior arch of the notochord; or, in other words, it is the true sense to the hæmal arch.

In introducing to the notice of the reader the sense apparatus pertaining to animal, as distinct from vegetable life, a totally new order of membranous differentiation occurs, and general metamorphosis; the true transitionary membrane being the mammary. And here a few general remarks will be made.

For the sake of uniformity, and perhaps also to make the subject more simple, the plan of giving the three membranes —serous, mucous, and contractile—will be adhered to; and the fourth division, a true animal *nerve membrane*, which includes sensation and motion, will be given as a sense for which a tripartite membrane is supplied, to enable it to fulfil its functions, and to obey the mandates of its requirements, which proceed from the brain, or great emporium of all the senses. But, rigidly speaking, the membrane of *animal life* is a *quaternary membrane*, or a *quadruple membrane*, and consists of nerves of sense and motion of a voluntary character, and also of an involuntary character, one proceeding from the spinal cord, and the other from the brain; and, in speaking of animal life membranes, they ought to be expressed as each membrane consisting of four divisions— nerve, contractile, serous, and mucous membrane. But, as before said, nerve will always be understood, and each membrane will be spoken of as a tripartite sense apparatus.

So each special animal membrane is, as it were, nothing more nor less than a series of mechanical appliances, here called sense apparatus, placed at the disposal of each special sense, to enable it to accomplish that fixed purpose in the animal economy for which the sense was destined, either for protecting and preserving the being, or to make life more enjoyable; whilst the brain and spinal cord are membranes to themselves, ruling and co-ordinating the several sense membranes or sense apparatuses.

Moreover, in the sense apparatus, membrane undergoes an entire change in differentiation in its animal life. The order of development is essentially *segmental*. To illustrate this matter, let it be supposed that the typical vegetable membrane shall be the intestinal canal from mouth to anus. We have the serous and mucous membranes tolerably continuous, and also the contractile membrane. The reverse of this pertains to the sense apparatus of want, and also of force. Take the former. The small intestines, liver, heart, and lungs, as well as stomach, claim some share in the pneumogastric, though, for reasons longer than can be here expressed, the liver and small bowels are provided to some extent with nerves of force as well as of want, or the hæmal sense; but accept its value as moderately correct, we have the chest and abdomen under the guardian care of the hæmal sense. This sense has for its tripartite arrangement three distinct membranes—serous, mucous, and contractile.

Of these, take the contractile membrane; it includes all the respiratory muscles—as intercostal, abdominal, and diaphragm, etc. Though all working in unison to one end—namely, to promote the function of respiration, and more occasionally to empty abdominal fulness in the actions of vomiting and defecation—yet, in respect of attachment, as of origin and insertion of muscles, though the end is emptying out to make room for healthy material, with the exception of the diaphragm, every part of the contractile membrane is isolated by origin and insertion, as so many segmentations, under the title of *muscles*.

As of the sense apparatus of the hæmal sense, so also of the sense apparatus of the sense of force; it is essentially segmental in the contractile membrane. Whilst all the membranes of animal life are essentially segmental, that of the serous membrane is pre-eminently so in the *sense of touch*. But of that more by-and-bye. To express the same view in another form—instead of a continuous membrane, the same membrane is repeated over and over again in completing any proper or special membrane.

To give a simile of a quadruple membrane may perhaps assist our understanding of the engrafting of the senses, on leaving the brain and spinal cord, as being a part of a special membrane external to the brain, and henceforth incorporated with the functions and working of a distinct apparatus, whilst in its major function it is debtor to the brain itself.

Let a gasometer stand for the brain; the main pipes and service pipes for the spinal cord and roots of nerves. But from the house meter and within, all gas apparatus and piping inside the house is bought and sold with the house; but though the proprietary is distinct, yet all burners in a house are more dependent upon the supply from the gasometer forty yards off, or one or two miles, than they are for any supply they get as the result of gas-fittings within the house. This is a rough simile, but it may serve to illustrate what is meant by the senses in their ultimate distribution outside the brain and spinal cord; when external to these centres each sense has its own tripartite membrane, as each house has its own gas-fittings distinct from the gas-works.

We now go on to consider the hæmal sense, or par vagum, *alias* the sense of want; as hunger, etc., it occupies its proper arch, and has its own tripartite membrane—serous, muscular, and mucous.

The ribs, sternum, and cartilages represent the *mucous membrane*, the intercostal muscles, and the diaphragm, with the abdominal and perineal muscles changed and modified, with the abdominal ribs arrested in development, so as to allow freer motion of the trunk upon the limbs; these all combined represent the muscular or *contractile membrane;* and the *synovial sacs* placed at the ends of the ribs represent the *serous membrane*. Segmentation in the serous membrane at the heads of ribs is here represented in a very complete form; with the attempt in the sternum, in man and most mammalia, to form one continuous portion of bony membrane, *alias* mucous membrane.

Again, as contrasted with the interosseous muscles, the various directions and forms of distribution of the abdominal muscles must be viewed as a species of muscular segmentation.

So that one of the first indications of this sense apparatus is *segmentation*, which must be viewed as a kind of differentiation in the main, distinct from that of the differentiation in organic life; arising from the fact that membrane repeats itself in the form of continual segmentation in all its tripartite divisions. Again, in the great change in differentiation of the several divisions, the contractile membrane and the mucous membrane—one as the *striped muscle* of animal life, the other *as bone and cartilage*, and the third very little removed in cell differentiation, as synovial membrane, from serous membrane in the vegetative organs of the body—the principle of segmentation, which appears to be abhorred in the one, is common in the other, and this makes the great difference between them.

The apparatus for protecting and defending the hæmal sense, as in a castle, from external injury, especially in its distribution over organs which bear external pressure badly

—as the heart, lungs, etc., the liver also—is very singularly protected by the bony framework; whilst, on the other hand, this apparatus is made subservient in its offices to the next great sensient apparatus—the tripartite membrane of the apparatus of the sense of Force.*

This apparatus may be justly termed the *complete subdermal sense* apparatus, and consists of the limbs and jaws, and the vertebral column; and the mammary, as the *transitional* subdermal membrane.

In this the osseous framework stands for the mucous membrane, synovial membrane for the serous, and striped muscle for the contractile membrane.

Before entering into an outline of the relations which this system of bony framework stands to the sense of force, it may be well to state that it was this tripartite membrane which first led to the hypothesis that the human frame, and animal life generally, consisted of a *succession* of tripartite membranes—the lower in the scale of life the fewer, and the higher the greater number and complexity of tripartite and metamorphosed membranes of every conceivable form of cell differentiation.

Whenever there were good grounds for believing that there was diseased bone, as *caries*, there was a distinct form of border and eversion of the ulcer on the skin opening opposite, or more distantly placed, but leading to the diseased part; when there was no *apparent* pus for weeks from diseased bone, and long before it became loose, this ulcer on the skin, with the everted edges, told one tale—namely, diseased bone was near.

Why, it was asked, should the skin have such a sympathy with bone? What is there in bone distinct to other tissues? It is often the case that matter partly escapes from a cellular abscess beneath the skin, and long before all has drained away the outward opening in the skin heals, only to reopen

---

* See *Monthly Journal of Medicine* for February, March, and April, 1855.

at that or some neighbouring spot, when the matter has, like leaven, in neighbouring structures turned more to matter; but in bone no such thing. The annoying little speck of bone has not been removed, and though for one or two days the outer ulcer on the skin *appears* as if it would heal, yet month passes over by month, till at last the dead portion comes away; and not till then will it heal, and that soundly in a day or two.

The question recurred and recurred, again and again, Why so much sympathy for bone in the integument, as compared with other structures?

Compound fractures illustrated the same matter, but then, as the late Mr. Skey has said in the *Lancet* for August, 1870, that the bruising and contusion of the surrounding parts in the fractured bone lowered vital integrity, and so greatly prolonged recovery. But a case occurred in 1865 that set this view at rest, and carried with it a strong conviction that between the integument and bone there was some basis of sympathy more than we yet fully understood.

The same had been observed in *many other diseases* of a chronic nature, the sympathies of which were more deeply seated than anything the nervous system could explain, and of a more decidedly vital character.

But, without further entering upon practical and pathological matters, the case will be briefly given.

F. B., æt. 34, in February, 1865, had a furunculus in the right thigh, about the middle, from which in due time a core came away, and then healed. In May he had a second one higher up, and somewhat below the trochanter major, but more posteriorly; from this a large core came, and was in process of healing, and he at work again, though in removing furniture he left the heavier work to his men.

Before it was entirely healed, and when removing furniture, he gave a helping hand to some very heavy furniture, and in the strain felt a snap in the right thigh; it was at right angles to the long axis of the femur, and close to its lower third. He walked home with much difficulty, and was surprised at the trouble he had to get upstairs to bed; but as there was very slight displacement, the weight of the body for so serious an accident, was well borne.

Examination proved the nature of the accident, and when recovered, the thick, hard callus settled it still more, if more could be required. It was seen later on by a surgeon of much eminence belonging to one of our large metropolitan hospitals, who also confirmed the diagnosis.

Here there was no bruising or contusion accompanying the fracture. The fracture was clear and unmistakable. The two preceding furunculi were also unmistakable, the last about ten inches from the seat of fracture.

If air getting to the connective tissue is the chief cause, as some maintain, of matter forming at a distant part from the opening, there had been a fair opportunity in the opening in the integument caused by two previous furunculi, for inflammation and abscess; the last of which, at the time of fracture, was fast healing.

But the fracture itself, contrary to all that might be expected, proved as intractable as any compound fracture is wont to be. By the third day the tumefaction was very great, soon to be followed by an enormous and long-continued discharge of pus, first by the opening in the skin made by the core of the furunculus, and then in three separate places—one near the knee, one near the fracture, and a third opening was about five inches above the seat of injury.

In six months the patient walked moderately well on crutches, and in three months more with a stick, and returned to his usual duties after that; but the last-mentioned opening remained open, and, extending by a long sinus to the callous bony structure, for more than six years, discharging daily a small quantity of yellow, and sometimes a chalky-looking matter.

Air, it was certain, could not be the cause of such a tremendous disturbance in the fractured limb *per se*, because the opening in the skin had been before the fracture, and twice in the space of three months.

The bad condition of the system could not be the cause, because it was the same as existed immediately before the fracture.

Bruising and contusion could not be the cause, since the limbs had had no bruise, or blow, or injury from without whatever.

The solution to the difficulty was one that had long presented itself, not only with regard to bone, but other structures. Could nerves account for it? This had long been doubted, but here no doubt was left. The sympathy was of a morbid nutritive character, and prevented assimilation through the entire connective tissue, which appeared to run parallel with two distinct tissues—bone and integument.

The inference drawn was that there was a closer identity of origin in two such opposite structures than was usually held; and, carefully weighing the morphological doctrines of Goëthe, the conclusion drawn was that bone, in its very highly-organized condition, was very distinct from a mere hard mechanical substance, and the constant supply of blood and cell changes known to occur in bone indicated in the midst of pure mechanical agency active vital changes, coupled with peculiar cell secretive power, that entirely excluded bone from being viewed as a simple mechanical structure; and that probably bone acted, as by deputy, *through the blood*, in using lime as an antiseptic, as well as by its agency aiding the process of disintegration in other structures.

Again, examining skin and bone with muscle and serous membrane, a general conclusion was arrived at—namely, that all complete membranes, howsoever modified in structure and cell differentiation, yet retained three primary elements.

1st : A contractile membrane.

2nd : A secretive and assimilative membrane ; and

3rd : A membrane, in all instances devoted more or less to mechanical purposes.

As the subdermal membrane is the largest and most elaborate of the tripartite membranes, and that one which led to the analysis of the rest of the structures of the body as so many varied and highly metamorphosed membranes, the grounds for suspecting so singular an arrangement of tissues and textures, as that based upon a very complex scale of tripartite division of membrane have been here, as it were, introduced, that a notion might be given of what were the lines of reasoning, from so small and unusual an incident as a compound fractured femur without a blow or a crush, which led to the present tripartite membranous theory.

No attempt at a refined or very careful analysis of the tripartite membrane of the *sense of force* can be given, because it would consume space equal to more volumes than one, if carefully and minutely examined.

When a man enters a boat to row himself on the water the motion of the boat is reversed to his sight, and he has to move backwards instead of forwards, because of the mechanism of his own frame, plus that of the fluid in which propulsion is effected ; yet a fish moves head foremost and a fowl does the same in air or water, and so do all vertebræ and the entire of insect life.

This arises from a singular principle of mechanism belonging to vital mechanics, called, for want of a better name, *evergency*. The name, it is admitted, does not appear to imply any principle of mechanism whatever ; but motion in the universe appears to be based upon some common law or principle, which fails not to assert its pre-eminence, even in little and trivial matters. But the leaning to *grow outwards* and to *move outwards* has a wide-spread application; but a law or principle of mechanism in Nature was first suggested by the fact that many mechanisms of man in their grand finale had a backward motion, but in animals, though capable of backing, yet their natural motion was forward, or in the direction of the head. The conclusion arrived at was that the *combined actions of muscles* turned towards some one particular *axis*, or point.

The direction of fish, in their combined muscular motion, is centred towards the head; birds near the sacrum, or near its juncture with the vertebræ; and mammalia towards the seventh cervical vertebra, but for man a double axis is claimed, one at the sacro-lumbar articulation, and the other at the seventh cervical.

Of course, there is no intention of giving a laboured account of the numerous data and details used in coming to such a conclusion. Neither is there any request made that the conclusions may be accepted as correct, because no reasons are given.

The object of giving these conclusions is merely to facilitate description, and that one point may be steadily kept in view—namely, that gravity and resistance are the leading elements for consideration in viewing the greater part of the osseous and muscular system, as one comprehensive and complex tripartite membrane.

A fish has, properly speaking, but one limb—namely, the vertebræ and muscles posterior to the dorsal region, or region posterior to the ribs. The fins are as so many side sails to poise and adjust motion; but, to speak in vernacular language as distinct from scientific, the tail has the major motion and the chief propelling motion, or the posterior limb of the fish is, so to speak, the only limb of the fish;

and with that limb is combined motion that propels the head in advance of the location in which the fish was, before such motion commenced.

In the motion of the fish the greatest power of resistance is towards the head, in the direction of the great premotionary sense sight, which sense, in relation to comparative or gradational anatomy, is next in order to the sense of want, or the par vagum.

Again, the greatest proportion of weight is anterior to the tail or posterior limb, so that propulsion has weight in advance of the propelling power.

Birds have the same peculiarity, for the greatest weight is anterior to the *axis* of their motion, or *combined result of muscular motion*. It is scarcely fair to mention it, but this was first arrived at by experiments, not upon birds, but upon one or two foolish Cheiropteræ, which were led astray by white traps. Till then no trouble existed as to the way of the eagle in the limpid air, neither had the Duke of Argyle written at that time upon this elegant and beautiful subject, and made it popular.

The experiments were conclusive and decisive. The upshot of the whole matter was this—wing action was so arranged, that force from the motion of air by muscle was backwards, and that it was greater than the gravity placed in front of it. Hence the head of the bird moved in advance so much every motion of the wings.

When birds swim, and when they walk, the gravity is greatest in front of the axis of muscular motion, which is near to the sacro-vertebral articulation; when walking the toes are anterior to the axis of gravity, as also in running; but the axis of muscular motion is between the joint (as of the hip or acetabulum) and the sacro-vertebral articulation, and very slight adjustments of the axis of gravitation determine whether in running, flying, swimming, or walking motion should be directed into its greatest or slowest form of speed—flight always requiring gravity in the greatest excess anteriorly to the axis of muscular motion, for which purpose the tail acts as a sure and ready rudder to give the balance of direction in relation to gravity.

As for mammalia, the seventh cervical is invariably lower in the body of the vertebra than the sacro-lumbar articulation, save, perhaps, in the giraffe and elephant, and may be the Baska horse (a beast of burden employed in Central Tartary), when moving on all fours.

But wherever the body of seventh cervical is higher than the sacro-lumbar articulation, there we have no power to jump in that animal, and it always moves either walking or running in the equilateral form—that is, one side moving backwards and the other moving forward together as equilateral halves, and never alternately, as in pigs, rats, asses, horses, and antelopes, etc., etc.

Hence all mammalia move towards the point of greatest gravity, which is towards the body of the seventh cervical vertebra, at which point gravity tends both from the sacro-lumbar region and from the head and neck. The trunk of the elephant, and the long neck and head of the giraffe, and the heavy head of the Baska horse, all tend to compensate for the elevation of the body at the seventh cervical vertebra.

Of course, in these latter animals, the long withers or spinous processes stand for nothing. Elevation must only be taken from the body of the vertebræ, and not from the spinous processes.

The outcomings of this arrangement of muscle and bone, in a complex machine, is to enable force in muscles both to meet resistance in fluids, either of air or water, and also to enable the bony framework to maintain its balance during flight, swimming, diving, and rapid running, without fear of an upset or capsizing, to use a most expressive term; the apparatus of the sense of want, or par vagum, acting from a mechanical point of view, as well-stowed ballast, materially aids safe and rapid motion, and aids in giving a right direction to gravity.

Taking, then, the entire of this membrane as under the guidance of the sense of weight or force, its great object is locomotion; whether in flight, defence, or prehension, the grand point is one and the same—*locomotion*.

Its general outline is thus summed up—a vertebral portion, three sets of limbs, and three plants, upon which the limbs are placed or fixed, with their complement of voluntary muscles and synovial membranes.

Let it be granted that the occipital bone is in some measure the counterpart to the sacrum; that the *condyles* of the occiput are equal to the *sacro-lumbar articulation* of the os sacrum; that the *long tube*, posteriorly bounded by the spinous ridge of the sacrum, is represented superiorly by the *foramen magnum* of the occiput; that the basilar process of the occiput is equal to the anterior portion of the sacrum, and the posterior expanded plane or curve of the occiput, with its roughened transverse ridges, is equal to the posterior portion of the sacrum, with its spinous and transverse ridges—then we get to the entire terminus of the series of segmentation, known as vertebræ, with their expanded and modified terminations.

The plants, or bony attachments at either extremity of the spine, are highly modified homologies of each other, and subserve distinct ends in bone mechanism. The mastoid process, with its serrated articulation with the occiput; the zygomatic process, and its union with the malar bone, are viewed as so many modifications of the ilium, pubes and ischium, with limbs, fixed and modified to this occipito-pelvic plant; also the superior and inferior maxillaries, as limbs, but, as it were, in direction at right angles to the limbs of the sacro-pelvic plant. The teeth and dental apparatus are so many devices, which blend with the sense of force, that are parts of the integumentary membrane, and are supplied with very fine nerves of touch within, like to the digital apparatuses pertaining to the anterior and posterior limbs.

The scapula, with its supplementary clavicle, is the remaining plant or foundation upon which the limbs are set. Taking advantage of the hæmal arches, it is almost as highly modified in some of its homologies with the sacro-pelvic plant as the occipito-pelvic plant is with the sacral, and in all points is wonderfully adapted to avoid shocks, and from its muscular adjustments to obtain free and rapid motion with great strength.

Concerning the homologies, differentiations, and arrests of development that occur throughout the vertebrate series between the fore and hind extremities, it is out of the compass of this paper to supply; for only so much is given as is sufficient to direct the attention to the *extent* and *nature* of the tripartite membrane of the *apparatus of the sense of force or weight*.

To whatever intermediate function and use this membrane in the economy may be devoted, taken as an entire and complete membrane, its function and end is to interlock segment to segment, so as to secure locomotion in every form and variety in which we see it carried out in vertebrate animals.

It might be considered that here was a proper place to give an analysis of that most complete and thoroughly worked-out monograph, by Mr. Parker, F.R.S., upon "The Shoulder Girdle;" but, howsoever such an analysis might grace an humble paper of this kind, the tracing of one is to illustrate Nature's work in the order of progressive development; but the other is a more humble attempt to utilize the facts after they have been acquired, and to go no further, for fear, in wading too far in such delightful streams of knowledge, the understanding might get out of its proper depth.

Nothing need be said in vindication of the sense of force, as its discussion here would be out of its legitimate place; but it may be asked, Of what use, in contending with resistance of a directly physical nature, is a sense that gives a knowledge of the outside or superficies, as the sense of touch? Surely, if there is no sense to give knowledge of gravity or resistance, we have a large locomotive apparatus left without a guide to inform us of the amount or degree of force which should be used in opposing resistance; and force is purely an ideal inference, if strength and power are never felt in individual bodies or persons.

The tripartite nature of this membrane needs very little examination, as bone is viewed in this membrane as modified mucous membrane; the voluntary muscles belonging to the jaws, spine, and limbs, are the contractile membrane; and the synovial sacs of the joints, and the articular processes of the spine and condyles are the serous membrane—each membrane being marvellously segmented and adapted to the general end of locomotion.

The joints, and not improbably bone, *in a measure*, being supplied chiefly with the nerves belonging to the sense of force; their distribution to muscles is much doubted, for to get a true estimate of the degrees of contraction or force in muscles, it would require the most diffused and minute distribution of nerves conceivable; but if distributed in and about joints the amount of nerve tissue required would be very greatly reduced, with an equally good channel for measuring and adjusting force or weight, *since all force in some way or other* *has to be transmitted through joints*, and in them we have a ready medium, with a smaller area, to supply with nerve fibres.*

We now come to the third somatic sense, or the sense apparatus of touch, whose tripartite membrane is the integument.

This membrane is the most diversified of all membranes in its forms of morphology and modes of cell differentiation.

It consists of the contractile membrane, in the form of beaded muscle, and dartos; of a serous membrane, highly modified in the form of hair, scales, nails, claws, feathers, hoofs, etc., according to the special requirements in particular animals, birds, and fish. The basement membrane appears to be essentially of that character common to mucous membrane, rather than serous; and the perspiratory glands and oil glands are essentially of the mucous order, being inflections from the surface, and secreting defined chemical compounds, and here and there most conducive to the well-being of the animal. The serous membrane appears to be *in man* a kind of arrested and abortive membrane, and which is here called *fragmentary*. In the subdivision of the animal kingdom, fragmentary conditions of membrane are very common, merely serving one particular end, and no trace to be found elsewhere.

In man it is essentially fragmentary, and also in swine; but in all instances it is segmented in the highest degree, and then hypertrophied as in hair, feathers, scales, etc. In nails and hoofs it constitutes a terminal segment. Horns belong to the same category, and the spines or quills of the porcupine and the carapace of the tortoise, the armadillo, and numerous fraternity of the same outward hard casing. The

---

* "Upon Nerves distributed to and about Synovial Membrane, as being a Special Seat of Nerves of Force." See *Monthly Journal of Medicine*, February, 1855, Edinburgh.

teeth are not exceptional, but they contain within them the conditions of a serous and mucous membrane, or bony and enamel substances, etc.

It will be said that, in ordinary serous membrane, we have no ground for supposing it ever assumes a truly fine segmentation, so as to form hair, etc. But even here we have in the lining membrane of the ampullæ of the semicircular canals of the ear fine hair-like processes, which membrane is essentially a serous one.*

But membrane in morphology, from its very metamorphic changes in cell differentiation, must alter in its form and structure; but in its function it will always possess in every membrane the essential type in relation to the rest—namely, physical and mechanical convenience without either contractile or active vital function. It is even doubtful whether serous membrane could on its own surface form a true ulcer, like to bone and mucous membrane, on account of its peculiar vital endowments, though apparently so highly organized in man, as in the pleura and peritoneum.

It has been already remarked that nerves of force are distributed on the integument in certain parts, as in the bend of joints, the palms of the hands and soles of the feet, along the mesian line of the lips, centre of the forehead, the nipples, and between the fingers and toes. In all these places, if water is applied, and then blown upon for a few seconds, cold is very quickly felt.†

Touch, on the other hand, is found at the lips and palate generally, especially the tip of the tongue, and the pulp of the teeth, the finest particles being recognized when pressed between the teeth, and the tips of the fingers and toes.

---

* *Vide* Huxley's " Elementary Physiology," page 223. 1869.

† *Edinburgh Medical Journal* for March, 1859 : "An Experimental Inquiry into the Existence of a Sixth Sense."

The conjunctiva is also very well supplied with this sense, and possibly bone and muscle in a *very slight* degree, but bone more than muscle.

Touch is the true prehensile sense of the body, as located in the mouth or lips, the hands, and especially the tips of the fingers and toes.

But touch is something more than this. It has a unity of sensation that gives a kind of ubiquity of *feeling all over;* one part cannot well be touched without the entire body feeling a unity of pleasure or pain, of warmth or of coldness, of creeping or curdling, according as its action is excited by external sources, or by mental induction or impression.

In this general feeling, if it is brought into close collision on the integument with the sense of force, for we have with the impressions of softness, sharpness, smoothness, or warmth or coldness, associated the feelings of resistance, hardness, weight and burden, strength or vigour.

This blending and co-relation of senses is beautifully exemplified in the neural arch of the sense of force, which further illustrates the blending of force and touch in the spinal arches, and foramina, which subserve for the mutual distribution of both these senses, both centrally and peripherally, and, in a great measure, has been the cause why senses so distinct in function and office, are so usually confounded as one.

Though more directly relating to the brain, yet a word may be said upon the three somatic senses—the sense of want, of force, and of touch.

It is usual to consider the striped muscles as strictly voluntary, with a certain amount of unwilled action, or sustained excito-motory or spinal continuous action, *but for initiation* dependent upon the will.

Very little reflection is required to rectify this palpable error.

Encase—as was once done by an artist, in taking a cast of the chest of a pugilist—the entire thorax in a wall of plaster of Paris, and in much less time than an hour the sufferer would be dead from asphyxia.

For why? Because the expansion and contraction in the air cells of the lungs is the minimum of respiratory action, carried on by inflating and compressing the lungs, through the medium of unstriped muscle; for the intercostal and abdominal muscles, with the diaphragm, have by far the most to do with the respiratory function; and in the spinal cord the respiratory tract of Sir C. Bell is the chief ministrator of nerve tissue to these parts, but by no means the only one.

These muscles never cease, save by an effort of the will, to act day and night from year's end to year's end quite as much as the heart. Life hangs on the balance of this continuous action, and, so to speak, the major action is as involuntary as is the peristaltic action of the bowels.

Hence striped muscle is, under certain conditions, as uniform in its action, and as constant, as the beaded muscles of the vegetative organs; whilst those striped muscles belonging to the limbs, etc., take rest as much as the senses do during sleep. Of course, in muscular structure, in health there is, whether contracting or not, a certain amount of tenacity or vitality, which appears to always hold them in readiness for action; but, so far as direct contraction and relaxation are concerned, they are in sleep, and at many other times, motionless.

It is evident that the functions of locomotion and rest are so blended in these two senses, force and touch, the latter being the sense that most incites to rest—smoothness, calmness, and softness or repose—that the continuous action of one is adverse to the other, and co-ordination of functions, in relation to the securing an end, is essential.

From a mere animal point of view, the integrity of touch and force are dependent upon the conditions indicated by want for their continuous activity either in procuring food, or by exercise, increasing the demand of fresh supplies of air, etc.

Hence two senses are the servants, or handmaidens, of one—namely, the sense of *want*—the indications of which are material supplies for material wants, and material rest when those wants are supplied. Hence, again, the successive alternation of repose and exertion.

How are two such contrasting senses to be reconciled by subservient co-ordination and perfect reciprocity? Probably the cerebellum has more to do with this co-ordination than any other part of the nervous system.

The experiments of Dr. Dickinson, chiefly upon serpents, are much in favour of such a view.* In this important function it is not improbable that the cerebellum has a certain inhibitory power over continuous nerve action, which puts a bar upon either sense taking an independent and continuous course of action, but makes one to be in part dependent upon the other for its continuous and sustained action.

Not that this view *per se* excludes the cerebellum from having some nerve relation to the blood and assimilation. What are often called internal, or central fits, and which are frequently followed by changes in the blood, that tend to retard motion and free circulation, have, in all probability, some relation to a special function of the cerebellum which, in nerve tissue, almost classifies it with mucous membrane in relation to active vital function—a function which it can fulfil without injury to sensation, since it is not in direct relation to any sensuous function whatsoever.

---

* "On the Functions of the Cerebellum," by W. H. Dickinson, M.D., *British and Foreign Medico-Chirurgical Review*, Oct. 1865, p. 455.

As we have arrived towards the central nerve mass, the brain, which is the real seat of the senses, a general outline ought to be given of the metamorphoses and differentiation of this structure and its relation to the senses.

The question is put, What is brain? The brain is probably metamorphosed muscle or contractile membrane,* metamorphosed into an impressible *concentrating membrane*, in contradistinction to a contractile membrane.

Its proper serous membrane is the arachnoid, with the lining membrane of the ventricles, and its proper mucous membrane is represented in the spreading ossified membrane constituting the skull, segmented separately for each sense and in such a manner as to protect the delicate fabric or contents of all the senses in one centre of co-ordination or reciprocity, and hermetically sealed box.

The spinal cord, unless viewed as a mere distributive organ to the nerves from the brain, is but an external extension of brain influence, that saves the brain the necessity of continually applying itself to direct and adapt motion *once started;* for by a series of incident and reflex actions, it sustains the action *already started*, and, in so far as it continues action once started, it is of inferior order, or has a lower function to perform in the animal economy than the brain itself, to which it stands as clerk of the works during the master's bidding, and sometimes in his absence, or when the brain is having repose. Its bony segmentation is for convenience, and more equal nerve distribution.

It has a corresponding mucous and serous membrane to that of the brain itself.

The brain, moreover, is the centre of a series of limbs, whose special functions, by its being an impressible concentrating membrane, are to co-ordinate and direct motions by a

---

* *Vide* S. W. Mitchell, M.D., U.S.A., " On Injuries of Nerves,' &c.

species of interdependent necessary reflex actions, whereby it regulates and guides the actions of the several limbs into one harmonious whole of mutually subservient agents one to the other. The senses of sight, hearing, smell, and taste are merely a succession of suppressed limbs or extremities, whose object is better gained by saving material, and altering the mechanical contrivances, as so many instances of special morphology, so as to secure greater extension of limbs by superior and more refined forms of mechanical and, in taste and smell, chemical contrivances.

There is, perhaps, no part to which greater importance ought to be attached than to the four senses—smell, taste, hearing, and sight; they are truly the interpreters to the remaining three, which have already been given. They are, as it were, complementary senses, neither of which performs its functions well without the education of the other. Sight aids hearing, because sound is so much better appreciated when the head is turned conveniently for its reception; and sounds in their kind are so distinctive that, from experience, we know for a given sound we have a given object in form and size to look at, as a fiddle, a bassoon, a cock, or a crow to turn towards, before we direct our best attention to the point from whence it proceeds; but, after seen and followed by the eye, the sounds are more clearly defined and the intensity more accurately measured.

If so much can be said of sight directing and educating the ear, much more may be said of sound indicating the point towards which sight ought to be directed.

The ear is more universal in its appreciation of its own natural stimulus than the eye; hence sounds from behind, the sides, or in front are nearly equally appreciated, and in such manner sight is directed to the object from whence sound comes almost instanter. A horse trotting on a country road may be often heard half a mile off or a quarter of a mile, but it is nearly as well perceived by sound half a mile behind as half a mile in front; yet, if there were no hearing, the same horse would not be perceived till its rider brought it to the front.

Smell and taste are almost still more co-related in function and education to each other than even sight and hearing, and it is more than possible that part of the enjoyment of food receives the connoisseur's approbation through the combination of two distinct kinds of impression at one and the same moment.

Touch and force are so interlocked by the apparatus for their distribution, and perfect appreciation of the end which each have respectively to fulfil, that it is almost impossible, saving by special experiment, to isolate and distinguish one from the other, so mutually do they aid and reciprocate with each other in their respective functions.

But the sense of want, or of hunger and suffocation, are here represented as one sense, through the par vagum. Hunger being the chief want in man and mammalia to which it is directed; suffocation only comes into occasional action when placed in non-respirable air.

Hence, in its essential action it has no proper counterpoise or balance by another sense in direct relation to itself, and, of all the senses, it stands the highest in its adaptation to self-preservation; since food in one form or other is essential to existence, and in animal life prehension and locomotion, saving in the lowest forms of the sub-kingdom, are essential as a means to an end in procuring food. And, as we get into the vertebrate, sight and hearing appear more or less essential in directing and properly using locomotion and prehension; whilst smell and taste are the special sentinels attendant upon hunger, to guard against satiety being sought for, as by smell in hunting, or appropriated by taste, when it is not suitable aliment for the animal exposed to hunger. These sentinels are specially endowed with the proper appreciation of what is suitable food for most things likely to be placed within their reach, and is one of the forms in which instinct manifests itself.

In one way or the other, then, it may be said that the remaining senses are servants to one special sense, over whose door is written, in more than iron characters, "That self-preservation is the first law of nature." This sense is the sense or consciousness of hunger.

But in birds, as pigeons, rooks, swallows, etc., etc., and in insects, especially the winged insects, as the hymenoptera, it is probable that the par vagum is the seat of a true *double sense*, and in the sensorium gives two distinct kinds of impression or sensation. The power to retrace their habitations, when far from home, in such small animals as bees, and with eyes constructed upon such singular optical principles, cannot possibly be from sight, nor yet from smell, as they are far inferior in smell to the common *musca vomitoria*; but the distances to which they travel, and the certainty of their return, bespeak a guiding power distinct either from smell or sight.

Pigeons conveyed in hampers, and never once placed in a position of surveying the country across which they will fly in retracing their homes, yet, by winging their flight direct for their old abode, over one hundred miles, indicates a positive guiding sense. Whence do they obtain such knowledge of distance and locality?

It is presumed that the sense of suffocation, or impure air, acts only in man to show him the deficiency of oxygen, and in the asthmatic occasionally other telluric, or atmospheric peculiarities are included. In certain classes of insects and birds this sense is open to stimuli, connected with certain conditions of atmosphere, that are always in active operation with certain conditions of the surface of the ground, *or substratum*, beneath which special conditions are so far changing every few yards in continuous succession that, though they are totally unrecognized by us, yet, that to these winged tribes, they give distinct sensations in the act of respiration, which, reaching the sensorium, are there registered; and, when the desire for return arrives, then, one by one, the past order of sensations becomes only sufficiently intense to be agreeable to the creatures feeling, when they are returning within a given range from the point from whence they were first received.

This view makes sensation in its general principles, or binary product, a self-instructor by principles of contrast and comparison.

It will be said there are eight, and not seven, senses, and that previously it had been said that there was a decided objection to needlessly increasing the number; to which reply is made, that it is more than probable that there are only seven senses in any one body or being. The sense of smell is particularly deficient in most birds, especially such as arrive in fixed localities after travelling very long distances, as is the case with the cuckoo, the swallow, and the passenger pigeon.

The observations of any one individual are totally insufficient to settle a matter like the present, but the general notion entertained about the senses is this—that the *Telluric* or the *atmospheric sense* (for such a name may stand for the travelling sense), and the sense of smell, are mutually compensating senses, or where one is, there the other is absent, or for all practical purposes may be counted as absent. In man, for instance, it is so for a non-active sense for locality, that it may be doubted whether it has ever shown itself, saving here and there in some blind man, as Metcalf, unless perchance it is unduly exalted in asthmatics, when it is the result of abnormal action from disease, rather than that of normal action in health.

Again, in pairing off the senses, each pair may be divided into the active and reposing senses, though repose and action are here used merely as comparative terms. Smell and sight are those senses in mammalia chiefly used in the chase and pursuit of food.

The sense of force is the one chiefly engaged in locomotion, and moving the body from one place to another; and the sense of want or hunger is the initiatory or prime mover of all the rest, and rouses the other senses to action.

Taste, hearing, and tactile touch, with softness, smoothness, etc., are all more or less brought into use when the range of the active senses is applied to a limited area, and more bound by external conditions. An animal grazes whilst very slowly moving or standing, and man sits and eats. The notes of birds are given when near to each other, to apprise their fellows of distant danger seen, or else to soothe and cheer a mate whilst sitting upon the nest. Sustained notes appear to be rarely used during much motion or intense watching, so that the pleasurable indulgence, both in sounds and taste, is only known under circumstances of comparative stillness and inaction. Tactile touch is chiefly in action when more active locomotion has ceased.

So between all the senses there is a kind of comparative activity in one, and quietude in the other; whilst the sense of hunger, being due to certain states of blood, during repletion a complete negation is given for a while, to its activity, to be again brought into activity by fasting.

It is curious to observe that this one sense, having no proper couplet, and in wild animals is the proper initiative sense to put into action the rest, has in itself a kind of negative condition leading to rest more or less complete. For when satiety is attained hunger has no existence, and impressions from that source are null and void, and with that a general tendency to quietude and inaction.

If the seven senses are more carefully considered from another point of view, they naturally divide themselves into three body senses, or the somatic senses, and four *accessory or supplementary* senses, or the παραίτιος, or the paraitic senses.

The somatic senses are touch, weight, and hunger, and give us a knowledge of self, or supply to us the condition of consciousness known as the *ego*—or conscious *ego*.

The paraitic senses are supplementary to and servants of the somatic senses to a greater or less extent, and include the senses of sight, hearing, smell, and taste.

These last-named senses are viewed as having peculiar apparatuses of metamorphic sense differentiation, distinct from the differentiations of the *somatic senses* in many and important particulars.

The paraitic senses are, saving in the sense of taste, elongated limbs, having special apparatuses for communicating afar off the conditions and circumstances of nature, through fixed and definite media, so that all necessary practical knowledge is supplied of external surroundings, by a new device in adaptation of limbs or *indirect* tactile touch, external to and beyond the range of the somatic senses. But the conscious ego (formed by the somatic senses) would be little else than a cipher in the world, with all its powers of locomotion, prehension, and satiety, unaided by the accessory senses, or servants to the somatic ones, yet with their aid the servants will carry out more completely what by himself the master cannot accomplish; and one of the main points that servants can do for a master, which the master alone cannot do, is, that through *representative servants* he can be in two or three places at one and the same time; so we can both hear, see, and smell at one and the same time, and the conscious ego is aware that it participates in all these several advantages by a kind of plurality of ubiquity.

This plurality of ubiquity in the paraitic senses, by instructing the somatic senses, is accomplished through the brain; and the brain, in function, is nothing else than the medium of union and greater extension of limb development, under a peculiar metamorphic differentiation, capable of adapting these limbs, or special senses, to higher and more extensive ranges of knowledge and perception.

Instead of giving a minute detail and defence of this singular view of the senses, here designated the *paraitic senses*, a remarkably short and tabulated view of the tripartite arrangement of membrane in the several paraitic senses will be given, with scarcely an explanation or defence throughout, as an outline forbids lengthened discussion.

I.

TRIPARTITE MEMBRANE OF THE SENSE OF TASTE IN MAN.

Mucous Membrane.—Mucous membrane of the tongue and salivary glands, and hyoid bone.

Serous Membrane.—Vocal cords and epiglottis.

Muscular Membrane.—Constrictors of the pharynx, and muscles moving the tongue, as the glosso-hyoid and glosso-pharyngeal muscles, etc.

II.

THE TRIPARTITE MEMBRANE OF THE SENSE OF SMELL.

Mucous Membrane.—The Schneiderian membrane, the lachrymal apparatus, and the lining membrane of the Eustachian tube and Tympanum, the two latter divisions being displaced portions of this membrane; the bony plates, as the turbinate bones, the vomer, the os nasi, and the Palati bones.

Serous Membrane.—The fibro-cartilages of the nose, and the cartilaginous rings of the trachea.

Muscular Membrane.—Levator palati, tensor palati, tensor tarsi, and the muscles comprising the external nasal group.

III.

THE SENSE OF HEARING.

This sense, and that of sight, appear to have *two tripartite membranes* — an essential, and a supplementary tripartite membrane.

Essential Tripartite Membrane, or Inner Ear.

Mucous Membrane.—Cochlea, with its modiolus and the semi-circular canals.

Serous Membrane.—Lining membrane of cochlea and semi-circular canals and otolithes.

Muscular Membrane. — Arytænoid muscles displaced to regulate the voice.

Supplementary Tripartite Membrane, or External Appendages.

Mucous Membrane.—The small bones of the Tympanum, with the membrane round the fenestra ovalis, and membrana tympani, being a species of basement membrane.*

Serous Membrane or Differentiation.—The cartilage of the concha as far as to the membrana tympani, and the elastic tissue of the Eustachian tube.

Muscular Membrane.—The muscle moving the small bones of the ear and concha.

---

* Of course, in fishes, where the small bones of the ear are differentiated into large lateral bones of the head, their associations with hearing are *nil*, and the impressions are chiefly given to the sense of force; but these bones really belong to the outer protecting department of the skull more than to any particular function of the ear direct, for this sense in fishes is feebly developed.

IV.

THE SENSE OF SIGHT.

The Essential Tripartite Membrane.

Mucous Membrane.—Membrana pigmenti, *alias* Choroid membrane.

Serous Membrane.—Jacob's membrane.

Muscular Membrane.—The ciliary muscle and ligament, the latter being probably modified muscle.

Supplementary Tripartite Membrane.

Mucous Membrane highly modified.—Cornea.

The Muscular Membrane.—The oblique and recti muscles, with their membranous expansion, the sclerotic coat; the orbicularis palpebrarum and the levator palpebræ; and the corrugator supercilii displaced.

Serous Membrane.—Crystalline lens, and aqueous and vitreous humours.

The highly modified and differentiated condition of the paraitic senses, and proneness to displacement *inter se*, renders it extremely difficult to trace their relations and real locality, since each in its function is so much aided by one in close relation to it. And as in the brain some appear to combine, as it were, by commissure, and to blend the products of their various stimuli by a peculiar interlacing of fibres reaching to certain cineritious cells or grey substances, so, in their external apparatuses the tripartite membranes of these senses appear to interchange one with another.

As the sense of smell supplies to the eyes and ears a certain amount of mucous membrane, as in the lachrymal apparatus, and to the tympanum and Eustachian tube in the ear, so the sense of taste supplies, in its serous membrane, the arytenoid cartilage and epiglottis, which are regulated by muscles in relation to the muscular membrane of the internal ear.

If the sense of smell includes the nasal apparatus, down to the trachea and larger bronchi, as its proper serous membrane—the mucous membrane, as extending to the same parts and to the eyes, as the conjunctival membrane, and to the ear, supplying, by displacement, the Eustachian tube; and the nasal and palatal muscles be viewed as the contractile membrane—then we have in the sense of smell a certain co-ordination of structures and apparatus which, in the economy of Nature, is closely interlinked with mechanism and function.

Again, as we descend in the animal scale, we find fishes merely selecting suitable localities for spawning, and the male and female nidus of a new generation requires no further intercourse than that the myriads of ova cast in the water should, after extrusion, be fructified by the milt of the male. Between these two stand an intermediate group, the reptilia, which are passed by for the present, with their reduced vertebræ in the cervical region.

Here, then, we have a distinct line drawn as to the order of procreation, and the amount of care necessary to bestow upon the young offspring. The lower the scale, the less need of care for the offspring. In other words, with this lowered standard of care for the young, runs *the lower standard of animal heat*.

If to this be added, as we lose the sense of smell, and with it the accompanying sense apparatus, in the trachea and larger bronchi—that we have an imperfect ear, so that no Eustachian tube is wanted; and an eye of great perfection, but from the medium in which it lives no lachrymal apparatus is wanted—then we can come to quietly consider how, in such general outward conditions, *a bony mechanism should be withdrawn*, and a distinct class of vertebrata should present themselves under a great variety of forms and sizes, which no longer need a distinct bony region. This bony mechanism is the proper *cervical region;* and the shoulder girdle is here, but a supplement to the back of the head; and flanking between the shoulder girdle and the maxillaries are the small bones of the ear in man and mammalia, enlarged to an enormous size, as a kind of outside hoarding, to aid, with the shoulder girdle, in protecting the heart and gills in their newly-acquired position and function outside the ribs; because the sense of smell in water is useless; to which all other parts are adapted, as the absence of feet and legs, for without a neck the others would be useless.

In such a brief outline it is impossible to go further than indicate the effect of removal of a tripartite apparatus in altering the whole phase of sequences. Neither can it be said that such phenomena as neckless vertebrata are a necessity from the medium in which they live, and the velocity with which they have to move, either for food or for safety. For though their own condition is necessarily a water one, yet we have tenants of the deep, and tenants of the rivers and lakes, which run from fish to reptiles of varied size and form, and even to mammalia themselves, which, though beautifully metamorphosed and differentiated —so as to be adapted to the element in which they mostly, and some entirely, live—yet they possess some kind of a cervical region, and with it some abortive attempt to develop the sense of smell. But in most birds and insects a modified trachea and air cavities and vessels are used as an apparatus for the *Telluric* sense, *which is a complementary sense to that of the sense of smell.*

We will now leave the subject of the senses to shortly refer to that organ, the centre of all the senses, and the source from which all active animal motions tend, or from which they originate—namely, the brain. As already indicated, the brain is here viewed as metamorphosed muscle.

It may be asked, How was it that the brain came to be adjudged as metamorphosed muscle?

To give the real reasons in full is almost impossible, as one suggestion came from one source, and another from an opposite source; and through a long series of years suggestion after suggestion will arise, which, when all are collated, lead to some more or less general conclusion. This again wants sifting and resifting, for the purpose of exhuming the dead matter and retaining the living and true matter.

For many years a general notion had existed, as before stated, that our planetary system, and a very extensive amount of our vital mechanics, were based upon a system of *evergency*. The principle is so far carried out as to explain or account for animals being organized and adjusted, that, as symmetrical bodies, they are stronger on one side than the other, or possess the greatest precision of motion on the right side (which is, of course, most easily proved in man); and that man, for instance, with the planets, in the order of his motion, is in perfect harmony when naturally endowed, or when he is right-handed, with this general principle of evergency.

For, let a man revolve upon his heel and follow his nose, and not *recede* from it and go backwards way; but let him go forward and follow his nose—which, by-the-bye, is rather a homely form of expression—he will invariably move in the direction of the earth's axis, and *the most active* or right side will be placed outwardly, and he will revolve on the left foot from west to east, in the same direction in which the matter of the earth has its greatest and least acceleration.

In other words, the side of strength will be *outwards*, and of weakness towards the centre of revolution, which is *inwards*.

The principle, then, of evergency, or turning outwards—as vegetables, which grow by the root much less than outwardly from the earth—is apparently implanted upon nature generally, with here and there exceptions; and it is the same principle which is implanted upon all the senses. For in their bond of mutual recognition or brain-consciousness, the sense apparatus, in all, is external to the centre storehouse or emporium of consciousness.

Observing, then, that the emporium or brain itself reflects the entire product of all the senses by an impressible power, which, as by a looking-glass, exactly duplicated the external recognizers, or sense apparatus or limbs, it was inferred that that principle of duplication must be the true and exact counterpart to *evergency;* and as a consequence the principle of conscious reflexion, or re-duplication of the senses in the form of consciousness—we say, as a true counterpart to evergency—the brain function which stamps itself by the power or principle of consciousness, is correctly counterparted by the principle of *invergency.*

As, then, muscle and tendon are under the regulation and direction of the senses, and as the senses peripherally and centrally are specially modified to receive impressions from special stimuli—as air, colours, odours, etc., etc.—and by tubing convey impressions from one to the other, as from the periphery to the centre, according to fixed laws of reciprocity, so in that reciprocity *from the centre*, metamorphosed muscle or brain is identified to be the chief agent of direct sympathy between muscles and the senses (and also from the spinal cord); therefore, between these tissues there must be a close identity of structural origin, as they have such an entire identity in their aiding, carrying out, and perfecting each other's functions.

It is therefore inferred that white conducting fibres are analogues of the tendons and the sarcolemma of muscles;

and the ultimate nerve distributions in the several senses, as well as the cineritious matter of the brain and spinal cord, are analogues of muscular fibrillæ or true muscular structure. Thus the brain and spinal nerves, with their ultimate nerve distribution, are considered in their differentiations as morphologies of muscle.

Though it may seem strange, yet the muscles of voluntary motion are viewed as differentiated brain, responding to the cineritious or grey matter of the brain and spinal cord ; and, to speak in a figure, are a species of conscious or reciprocating agents to the brain, outside the domain of its own special conscious agency.

Again, for the brain to be truly invergent, the right ought to change for the left, which is the case, the left side of the brain ruling the right side of the body.

It is with extreme regret that, in speaking of the senses of force, touch, and want, no better authority can be given, so far as the writer knows, than those contained in the *Edinburgh Monthly Journal of Medicine* for 1855, including the numbers of February, March, and April, the title of the article being " Upon an Experimental Inquiry into the Existence of a Sixth Sense ; " and in 1859, for March, April, May, some further researches, including suggestions in relation to a seventh sense, in the *Edinburgh Medical Journal*, by the same author.

Whether from an over-weaning fondness for one's own child, or from a fair and honest examination of the contents of these articles, the writer will not attempt to decide; but length of time has rather confirmed him in his original researches, than led him to doubt or abandon them. For instance, in the sub-kingdom we naturally ask, Of what good would the sense of touch be to the common crab, save in the fragmentary form of the antennæ, with its hard and stony shell ? But force to such a creature would be everything, and balance, which goes with force. Again, Of what use would force be, saving to the foot, to such creatures as most of the molluscous animals?

But, in a crude essay like the present, details must be curtailed and broad outlines alone be suggested or touched upon.

In the vegetable kingdom morphology is already, at the hands of Goëthe, prepared to hand complete and exact, and not as the present animal morphology, in a most sketchy and hurried manner, with scarcely a moment's lingering to see the fields, whether they yield oats or wheat; the sketch is so rapid, and withal so remarkably defective as a treatise to itself; but it is only given as a kind of suggestive ideal system, which has been passed over in rapid succession, in order that in a bird's-eye view the whole of the animal kingdom might be taken in at one sweep. And now briefly for the vegetable kingdom.

The morphology of the leaf, through the flower to the fruit, is now left just as it is found (discretion being the better part of valour); so is the cortex, or the outer bark; but there are two things worthy of attention—firstly, the relation of the outer bark, or cortex, to the leaf; and secondly, the density of the wood in relation to the leaf.

1st: In such trees as the oak, the ash, the beech, and the elm, etc., all being simple leaves, the roughness of the bark, when the trunk is considerable—say of twenty years' growth —appears to depend, in some measure, upon the depth of sinuosity, or the depth of serration, or dentation of the leaf. The oak is deeply sinuous, and the bark is very rough; the hawthorn not far different, and it is very rough when growing *alone*, and where the trunk has stood twenty or thirty years. The leaf of the beech is very even all round the margin, and the bark is the same; the elm leaf is *finely* serrated, and the roughness of the bark is very equal, but not deep. So of different kinds of apple, pear, and plum trees of twenty to thirty years' growth, in a large orchard, some will be more dentated, or serrated, than others, and the bark will agree much with the condition of the leaf. There is one remarkable exception, and that is in the compound pinnated leaf of the common acacia (*Robina pseudacacia*), where the leaf is very smooth round its borders, and the bark is very deeply grooved and rough. Upon carefully examining this leaf, it appears almost as if the mid-rib was a continuance from the petiole, and the leaflets were really transverse or side-ribs, called *lateral nerves*, with the parenchyma interrupted or arrested in its development, and therefore formed of a succession of *apparently* real leaflets. If such should really be the case, then the deep rough fissuring of the bark only confirms the general statement here made.

Again, the roughness or smoothness of the bark is not the only point connected with the bark which is connected with the form of the leaf; for, in some trees, the bark scales off in patches or flakes, and does not appear to groove in the line of the long axis of the trunk, nor at any angle with that far short of what might be termed, in rather freely-used language, at right angles to the long axis. *In such trees the leaves appear to be broader at right angles to the mid-rib than in the direction of the mid-rib*—as, for instance, the bark of the sycamore tree scales off in flakes and does not fissure; also the very singularly arranged bark, or *cellular integument* of the birch, which peels off in a circle, and yet is deeply fissured in the bark in old trees; but the leaf is well serrated on its borders, and is as broad transversely as in the line of the mid-rib.

Again, the stem as a whole, in the last year's shoots, or growth, is much more frequently found quadrangular in evergreens, or trees which, though not evergreens, yet retain their leaves till very late in the season, as the blackberry and the privet trees, or shrubs, than in those trees which shed their leaves in mid, or late in the autumn—as, say, in September and October in this country; but to this general statement there are many exceptions.

2nd: The density of wood, especially in its centre, when a tree is in its vigour—say, 40 to 60 years of age—and the *bark in the trunk is complete*, is, *cæteris paribus*, harder wood, especially in its centre, *in ratio to the thinness of the mid-rib, as compared with aggregate of the transverse ribs* (or lateral nerves). To illustrate this matter, a cherry tree in its centre is rarely ever hard wood when cut at 60 years of growth, though the intermediate wood is hard; the same may be said of the horse-chesnut. If, then, any of these trunks are compared with the beech or the oak at their centres, at 60 or 80 years of growth, the hardness is much greater in the latter than in the former. Then take an oak leaf and a beech-tree leaf, and examine the transverse ribs with the mid-rib, and the mid-rib is scarcely equal in the amount of its substance as compared with the transverse ribs; but in the cherry tree, and especially in the chesnut tree, the mid-rib is very large as compared with the transverse nerves.

Again, the size and length of the nerves to the mid-rib of the leaf, appears to have something to do with the length of the woody fibre, or with the fragility of the trunk or larger branches. This is well illustrated by two trees which develop timber in moderately equal ratios of time—namely, the beech and the elm; the beech forming timber in many localities where the other grows well, but the beech makes timber the quicker of the two. The beech leaf is much larger and longer altogether than the elm, and its timber is far less brittle, from the woody fibre being longer than that of the elm.

These observations are upon an extremely limited scale, yet they have included most of our forest and garden trees, and, when carefully examined, appear to be sufficiently uniform to claim a short notice, when passing in review morphology in its general bearings upon vital force, showing that there is a general interdependence of parts, in a whole, throughout all its general bearings.

But, as regards the approximation of identity in relation to the two kingdoms, there is but one general impression here maintained, and that is, that in cell development, throughout its endless morphologies and differentiations, there is nothing that really stands equivalent to a tripartite membrane, or any cell possessing a really genuine contractile power, self-existent in the tissues of the vegetable kingdom. All assimilations to contractile tissues are all so many modes of utilizing external nature, so as to bring about peculiar kinds of *mechanism;* such as the sensitive plant, Venus's fly-trap, the opening by night or closing by night of certain flowers, and the twisting of leaves to moist wind and exposing their stomata, etc., etc. The rising of the sap is due alike to the expansive effects of heat, and, may be also, to chemical changes, effected by the chemical rays of the sun, both aiding to compel a vacuum, Nature's great horror; and, for fear of falling into any quagmires from this source, she has secured a most perfect mechanical parasite, if such a term may be generalized, in her great agent, atmospheric pressure; add again to this capillary attraction, and the law of diffusion of fluids, especially aided by membrane, as shown by Graham and Duhamel in osmosis and dialysis, and then we have all the agents necessary to perpetuate successive changes and progress, in increased decay and in onward development, which are required after the first start of the true vegetable fructification of any special kind of ovum.

As there is no real contractile force in the vegetable

kingdom, there can be no real representative of a nervous system—above all, of a sentient nervous system, as the brain, etc., as is here maintained; for, as already advanced, the brain, which is part of a tripartite membrane, is viewed only as a peculiar morphology of muscle, and wherever there is animal muscle there we have some kind of a brain or neural mass, however small or oddly arranged; and these two systems, the cerebral and the animal muscular systems, are so arranged that, however fragmentary and apparently independent one system is of another, yet that these two systems are always mutually interdependent and co-related.

Yet, even here, vital force asserts its right to turn to use, in its multifarious products, the principles in one kingdom by adopting them in another; and the exogens and endogens are independent expressions of the principle of invergency and evergency in very complex mechanisms, side by side— very simple mechanisms, in which the different principles of inevergent and evergent mechanism are adopted. Thus we have, as exogens, the fine and complex mechanism from ovum to fruit of the massive oak, and the more simple one of a cowslip or a strawberry, etc.; or, of endogens, in the lofty cocoanut and the fragile grass, or the beautiful tulip. The principle is the same—its mode of application is as contrasting as a needle is to a steam-engine.

The general conclusion arrived at from the foregoing examination of matter, organized and unorganized, is, that vital force manifests itself obedient to general laws, and in detail adapts itself to particular mechanical principles, and to uniform plans of mechanical simplicity and complexity, in constructing bodies from the lowest forms to those of the highest order of organization.

The principles of mechanism, and the general laws to which the whole universe is subject, are so harmonious in their operations, that the Architect and Law-giver must be in perfect union, or have one mind and will; or if without mind and will, all has been done exactly as if mind and will were never absent.

We now turn to examine vital force in its more general and extensive forms of manifestation, as affecting large areas of the earth's surface; and the constant ebb and flow of those changes in vital manifestation known as Epidemics.

EPIDEMICS.

Vital action, or force, takes a much wider sway than that of merely retaining in integrity the component parts of a system, or interdependent portions of a complete whole or body, for it manifests itself in waves and patches over large areas and in particular parts of the world; and also its modifying powers are shown in long or short periods of endemic variations, and more widely-spreading changes, called *Epidemics;* or metamorphic conditions of disease, which are greatly concerned in developing the energies of races, and also in limiting the well-being of habitable tracts of the earth's surface.

In fact, coupled with duration, *change* is written upon every portion of the globe; and metamorphic phases of *disease and health* are spreading their special stamp of vital manifestation in every historic period; and each successive generation feels that change, more or less. The bare culture of the ground, the upturning of the soil, the destruction of many kinds of vegetable growth, to be succeeded by one, two, three, or more of a particular kind; the amount of forest, wooding, fencing, grazing, and cereal yields that any given track of land produces, as well as its manuring, watering, and drainage, all tend to regulate the amount of health, and the totality of disease and death in that given area. The food, labour, and exercise, the clothing, housing, and bedding of man and animals, alike promote disease or minister to health.

Certain localities, in their geographical features or geological positions, entail certain fixed and recognized diseases, as swampy lands. Lincolnshire and Kent, for instance, give (or once gave), as a birthright to their inhabitants, a large preponderance of ague. The Alpine and Derbyshire waters, being excessive in lime—the Alpine giving magnesia additionally—induce goître and cretinism. The ploughing or digging of certain kinds of soil, whether it be in Ireland, Scotland, England, China, India, or the Americas, etc., are frequently attended with special forms of fever and bowel affections, etc., etc.

Human tenements overcrowded, or long inhabited, without regard to daily cleanliness, and periodic renovations in the coating of the walls and ceilings, etc., and neglecting the careful removal of surface soil to suitable distances, and then to be subjected to special management, having as its basis complete or limited chemical changes—all, or several of these agencies, have, as their natural sequences, the power of inducing or intensifying special forms of disease.

Hence endemic disease, being disease in some special form, or of a more intense character in one locality than in another, embraces in its agency a variety of conditions, both personally and relatively to the individual or the community, as well as the incidental geographical tract, or geological stratum, in which a community or an individual may be located, which are inimical to health.

To these may be added, not only the moral and vicious habits of particular individuals, or even whole communities, but the particular hereditary vice or kinds of constitutions gendered by such habits and transmitted to their posterity, such as certain forms of dyspepsia, induced by epicureanism or 'gormandizing; unnatural thirst, as, when adult life is attained, from parents drinking ardent spirits, etc.; mental excitability and uncontrollableness of temper in the offspring, induced by over study and loss of sleep in a parent, or, what is often a much more common cause, the indulgence in youth of uncontrolled self-will and excessive liberty or power; habitual idleness and dilatory habits, entailing on the offspring, from want of food and force of example, harmless, or else cunning and inenergetic children, in whom systematic work is almost beyond the power of endurance. All these, and many more, resulting from moral defection, entail suffering, trial, and early-exhausted constitutions, which rob the children in man and woman-hood of health and energy to a great extent, and, if not actually shortening their days, in *utility* limiting their labour, to nearly half its proper and natural value, to themselves and their offspring.

The moral and vicious habits, in relation to health and disease, may not be improperly called entailed endemic or hereditary diseases; and in many respects they are the most inveterate of all diseases. They sap the vital powers, and are a leaven of disease, which damps and blights the primary vital impress continuously, and *more certainly* than any other endemic form of disease to which our race is subject, from the cradle to the grave.

If, therefore, treating upon the subject of epidemic and endemic diseases, a wider range is given than is usual, it must be borne in mind that this outline of epidemics, etc., has for its object the tracing out of external agencies in relation to vital power, and not the internal or inherent force which implants in matter a self-interdependent power, which is, in co-relation of elements and their interdependent functions, not only superior, but antagonistic to the surrounding inorganic world. Epidemic conditions are those external conditions, or continuous changes, occurring in the surrounding inorganic world, or the antagonism between higher and lower grades of organic life, which tend to reduce the products of vital power to a lower level, or to limit the number of individuals, or amount of matter that is directly under the sway of vital force.

From such a point of view it is evident a wider range of contingent agencies, always at work for evil or for good, in relation to the applying of a given force, as vital, to bring about a given result, must be taken into consideration in a general and comprehensive manner.

The sun, as the centre of our system, and the spring of all vital phenomena in the vegetable world, and from thence by induction to the animal world, requires a first consideration in examining the relation of the external world to vital phenomena.

Every part of the earth, during the circle of its journey round its own orbit, has exactly the same number of hours and minutes of light at the poles, the Equator, or the temperate regions; but, in relation to diurnal continuance of light, the greatest possible variation is attained, from twenty-four hours of exclusion of sunlight in the Arctic regions for a certain period of the year, and at another part of the year continuous sunlight for the same space of time— twenty-four hours—to the very moderate variation of an hour or less on either side of twelve hours in the Equatorial regions; and between these points every intermediate graduation of time in the twenty-four hours between the seasons of winter and summer. Therefore, *from the amount of sunlight* in the year upon every portion of the globe, no inference can be drawn as to health or sickness, barrenness or **fertility**.

Oblique and vertical rays of the sun much more materially affect the heat on the surface of the earth than the duration of light. The rays are most vertical in the tropical regions of the earth, and the least in the Arctic regions, where, on the contrary, the rays attain to their greatest obliquity. In summer the rays are always more vertical than in the winter or the cold and rainy seasons; *hence the direction of the rays of the sun is most important,* in relation to diurnal heat.

Humidity and high barometrical pressure affect the development of positive heat in the highest degree; hence valleys are hot with high barometrical pressure, and mountains are cold, where rarefaction is great, and barometrical pressure is diminished. And as positive heat, emanating from the rays of the sun, is a vivifying agent, and the chief of all known agents, it follows that long-extended plains, where rarefaction is very limited, and valleys well watered by rivers and streams, will give the largest per centage of vegetable and animal organization, and in equal latitudes high mountains will be most scanty in their yield of organized products.

But as year by year the same amount of sunlight, whether checked by clouds or not, is not the question; but the sun shines from above upon the same lands, and gives light year by year for the same length of time, and at the same period of the year, in every place alike throughout all ages—therefore, so far as relates to diurnal light, no part of the earth receives more one year than other, nor is the slightest amount of abatement to be found in solar distribution of light in any one part of the earth at the same period of time.

Spots upon the sun affect the magnetic currents on the earth somewhat, when present, but how this magnetic disturbance affects growth and vegetation at present we are in perfect ignorance, and among disturbing agents on the earth's surface, as yet, no positive inference of any kind can be drawn.

Our trade winds, siroccos, monsoons, etc., take their leading course of current, or direction of motion, from the combined action of rarefaction from the sun's heat, and the angle in relation to the earth's orbit at which the sun's rays fall upon any particular portion of the earth's surface; but if the appearance in time of such periodic wind currents depended *solely* upon the sun's rays, then they would always appear to the day and the minute in every part of the earth's surface where they are found; for the time the earth reaches any particular part of its orbit, and the amount of light which shines upon any particular part of the globe being always the same, the time of their appearance would be always identically the same, year by year, from age to age. But inasmuch as occasionally they are a few days too early, and frequently a week or two weeks later than their accustomed time, it is plain that some agency beyond that of the sun's rays has a powerful effect, and is a disturbing element in this perfect system of light administration, whereby the legitimate *effects* of solar calorification are limited, and diverted from their correct time of systematic recurrence.

The general inference, then, is this—that in relation to epidemics the sun may be viewed as an indeterminate agent, and a non-producer of epidemics. The one sole agency of the sun, external to his direct rays, as chemical, actine, calorific, and luminous, if such divisions of the rays are tenable in the present day, in the latter of which the solar spectrum apparently reveals identical elements in the sun to those that exist upon and constitute the chief elements of the earth, which in many respects, to say the least, is exceedingly problematic—to repeat, the sole agency beyond the rays of the sun are the spots of the sun.

Whatever may be the exact chemical constitution of the matter of which these spots are composed, two facts are apparently clear; first, that the spots are non-luminous;

secondly, that no rays proceed from the spots, but that they have a certain effect upon the magnetic, or the electromagnetic condition on the earth's surface.

This fact strongly confirms the views of Oersted, that the force called gravitation is perfectly expressed by the power or force called magnetism or electro-magnetism, the same as Lord Bacon suggested, as the controlling power which governed the moon in her circuit round the earth. For the spots on the sun, it is presumed, are masses of matter unsmelted by heat, and retain their inherent cohesion amidst the furnace by which they are surrounded; or, to say the least, they are free from the chemical changes to which surrounding matter is subject. Hence, in such case, the solid matter on the surface of the sun has a free and independent action, distinct from that of the incandescent surface, and in nature more akin to the solid noncandescent materials towards the centre of the sun; and, in their relation to our earth and the planets, give a truer transcript and interpretation of the nature of that union between the matter of the sun and the planets, from their relation to magnetism, than the incandescence of the matter of sun, ending in the phenomena of light, heat, and chemical changes on this earth, as well as in the sun itself.

Granting that the spots on the sun give us some key, *from the very slightly greater nearness* to the earth of solid *matter on the surface of the sun, to that of matter towards the centre of the sun*, yet if, in so delicate a medium of response to the slightest degrees of variation in intensity, the magnetic condition of the earth at certain points is affected and rendered apparent by the susceptibility of delicate electro-magnetic instruments, it only follows that the natural attractive power of electro-magnetism between the earth and the sun is slightly modified *by the proportion of distance* from the centre to the surface of the sun; and that the earth's surface, being unity, takes cognizance of the slightest variation of material change on the surface of the sun. But this natural sequence of perception of a force, in relation to distance, in two bodies mutually affected by distance, is real and determinable; but, in relation to material disturbance of matter on the surface of the globe as an efficient agent to promote disease, it is, so far as we yet know, perfectly innocuous, and its case must be dismissed as a true bill of cause to epidemic or endemic disease, as not proved.

Hence, as an active agent in promoting epidemic or endemic disease, the sun may be allowed to pass muster as being a non-active agent in initiating or producing epidemic or endemic disease, unless it be sun-stroke, the prickly heat of tropical climates, and liver affections, by, in part, suspending the functions of the lungs, as promoters of heat, and throwing a great plus of hydrogen and carbon on the liver, instead of excreting them by the lungs, as water and carbonic acid, when aided by the absorption of oxygen from the atmosphere.

Between earthquakes, volcanoes, and epidemics no direct co-relation can be established, since small-pox, influenza, plague, fevers, and cholera, etc., appear quite independent of these terrestrial disturbances. For instance, plague has affected certain countries and localities with great intensity, as Marseilles in 1720, Naples 1656, London 1664-6, Moscow 1770, Bassora, Persia, 1772, and many other places at particular times between these periods; but between the advent of earthquakes and volcanic eruptions, saving catarrh, as the mechanical effect of fine dust proceeding from the first outburst of Mount Hecla, there does not appear to be any regular coincidence between earthquakes and epidemics.

It is quite true that epidemics are constantly recurring, and so are earthquakes, but at very different parts in the earth, and by no means in a certain order, in relation to time, between one and another, so as to lead to an inference of cause and effect.

For instance, the year 1755, taken altogether, was as memorable a year as any in modern times for earthquakes. It began with Quito, in South America, and ended with the famous one of Lisbon; but this and the following year are not remarkable for any epidemic disease, nor the two years preceding the year 1755.

But Asia Minor was visited with very severe epidemic disease in 1760, which proved remarkably fatal. This country was, so far as is known, totally unaffected by the earthquake of 1755, which extended in a line from the north of Ireland to Morocco, Lisbon being the focus of its explosive powers.

In Mexico, in 1759, a mountain was raised by lava running from a volcano, but its distance from Syria, and all intermediate parts, presenting nothing peculiar in relation to epidemics, would lead to the general inference that the opening of this flue, to let the expansive force of heat have due vent, had nothing to do with the plague in Syria in 1760.

For the last hundred years and more, Chili has been a repository of earthquakes, and coast elevations have been considerable through their accompanying elevating force. In Concepcion is to be found a very focus of this oft-recurring force; yet in these regions yellow fever and cholera have neither been more severe, nor have they visited these regions and localities, nor yet any other epidemics, more frequently, or in greater intensity, than other parts of the North and South American Continents.

The history of ancient and modern times gives no better coincidence between epidemics, volcanoes, and earthquakes, than those incidentally given as a sample in the foregoing remarks. And identity of time, in relation to succession of events between earthquakes and epidemics (save in the cases of the Athenian Plague and the Levant), are so entirely deficient, as measured from this standard of order in succession of events, that either, standing in relation of cause and effect, must be entirely abandoned. Whether Comet Wine will meet with the same respect, the announcements of the wine merchants must settle; but their effects upon health belong entirely to the imaginations of superstitious people, and not to the close inductions of observing men; for, like earthquakes, the order of succession is equal to that of a cock crowing when a kite flies, or a balloon mounts the air—he may crow at such a time, or it is quite possible he may let it alone; but, generally speaking, neither would have a very close dependency upon the other.

Heavy falls of rain, snow, dry seasons, and over wet ones, also strong winds, each have an effect upon health, but it is purely endemic, and refers more to the effects of cold and damp, or over heat and exhaustion from defective appetite and intense thirst, which may be viewed as physical causes of disease in a great measure; but, added to heat or cold, direct want of food and of clothing, and over-crowding (the result of poverty), then we beget active local causes of disease, each providing their own morbific poison, as shown in typhoid or typhus fever, remittent fevers, dysentery and erysipelas, etc., etc., pneumonia, liver disease—as in the Irish famine of 1844 to 1846—and a host of minor affections, from hæmorrhoids to prurigo. But it cannot be denied that in camp and jail fevers—the effect of filth, over-crowding and defective nourishment—fevers have been gendered on the spot which have acquired a true infectious character, for all traces of their being imported appear to be totally incapable of demonstration, as shown in many of the visitations and reports of prisons given by John Howard; but numerous instances are given where prisoners brought to

Epidemics. 153

trial, after detention for a short time, have infected persons listening to their trial in open court.

Strong winds often have a most beneficial effect in producing an entire change of atmosphere, and introducing one free from the same morbific poison which the previous air tenant possessed; and by this means a disinfectant is supplied upon a large scale, and free of cost. Hence strong winds may be viewed as revivifying agents in the wheel of Nature, during certain periods of endemic and epidemic invasions, and stagnant air as a depressor of vital energy, and a reservoir for retaining in suspension morbific germs of disease. Cholera and scarlatina have frequently undergone partial suspense for a week or two, after strong gales, over any particular spot where these diseases are very prevalent; but the intermission is but for a season, and morbific poison soon re-asserts its prerogative, till time has allowed it to waste its native vigour, and to die a natural death.

When local causes are cited as feeding, and, in some instances, creating endemic diseases, such as fevers, dysentery, certain lung diseases, arising from given trades, as gunsmiths, needle and glass-grinders, paralysis from lead, etc.—and due weight ought to be given to these several agencies—yet in the midst of these local causes we find that one leading type of disease, as cholera, influenza, sweating sickness, or plague embraces all trades, localities, and countries—or in the course of a few years half a continent, or as in cholera, half the habitable portions of the globe are encompassed by it—then we have to appeal to some more general and universal cause or causes than such as pertain to trades, habits, vices, defective and bad food, over-crowding, bad water, bad drainage, and defective sanitary care.

Let infection, then, be the admitted basis of small-pox, cholera, influenza, yellow fever, leprosy, and plague. Granting these diseases have a more general and wider spread agency than can be conceded to any mere local cause, and, from receiving the title of endemic disease, are transferred to *the more extended field of epidemic disease*, the first point that attracts the attention is the selection which epidemics make out of the vast range of organized products; for if endemics select a limited area, or given trades, etc., epidemics select kinds of diseases, and special objects of attack, in such way and manner that one attribute of epidemics may be recognized as that of *isolation*.

It may be illustrated in the following manner. In 1771 mildew attacked the wheat crops in the United States of America and oats in Scotland, whereby these cereals, in their respective countries, yielded a most defective harvest. In 1830 to 1832, potatoes were considerably diseased in America, Germany, and Ireland, probably in the form of a fungi adhering to the fibrillæ or roots of the bulbs, occasioning a drying or withering of the bulb itself. But in 1845-6 Great Britain, Ireland, and a large portion of Europe suffered from severe epidemic potato disease, spreading over whole fields and countries with fearful rapidity, and as a deadly plague—the leaves and stems turning black and lifeless, and in another week or ten days the entire crops would become like charnel-houses of corruption and stench. This destructive disease still lingers in Europe, and in the Emerald Isle is a greater curse to the land than the plague of frogs in that charmed land of cabins and domestic stock.

What can be said of bulbs, cereals, smaller plants, and shrubs can scarcely be applied with propriety to vegetable organism of a larger growth, since disease may attack the flowers or fruit of vegetables of larger growth; but, as yet, no instance is recorded of the destruction of our larger trees, such as belong more especially to the forests and parks, as the oak, the elm, the beech, the cocoanut tree, mahogany or cedar trees.

Epidemics.

But, *upon the whole*, vegetable life, from the green sward in the field to the stately forest tree, enjoys an immunity which scarcely pertains to any class of animal life from insect life to the huge pachydermata, and, last of all, to man himself.

Preceding human epidemics, insects, gnats (and spiders in Germany, 1612, and in Spain, 1709), and locusts, with caterpillars in ancient and modern times, have swarmed and infested regions and tracts of land in a most destructive form, and frogs, etc., in a most unpleasant manner; but, in proportion as we admit their predominance in certain years, their disappearance in many instances indicates a greater blight to their onward procreation and increase than did their first development into a temporary pest, so that here again we meet with isolated increase and destruction almost in one and the same breath.

Epidemics have proved destructive to special classes of the animal kingdom, which indicates a power of destructive agency that at once excites curiosity, and claims an almost solemn reflection as to the guiding hand which can limit and determine over a widespread field such precise selection.

In Spain, in 1761, the dogs died in great abundance from some particular epidemic then prevalent. Their brother chips in zoological classification, but with tail and toes, that have created an invidious wall of separation, were brought into a kind of parallel approximation in the United States of America in 1771, when almost all the foxes died out.

No less singular was the isolation of the tenants of the oceans in 1529, for this year epidemic disease proved destructive to the porpoises in the Baltic. Fishes, lobsters, and oysters each in their turn have proved victims to pestilential disease. Birds, both domesticated and wild, have from time to time given evidence of widespread mortality in their ranks, as grouse, pheasants, pigeons, poultry, etc.

And if these, in their turns, are visited by epidemic disease, with how much more certainty can diseases in horned cattle, horses, sheep, and swine, etc., be cited as proving the fact of isolation, both in disease and species? The foot and mouth disease, since 1839, has from time to time attacked our horned cattle, but rarely the horse, if ever. The pleuro-pneumonia, which has swept thousands of milch cows from our sheds, has rarely touched the horse, save at the Cape, where horned cattle appear, until lately, to have been free from this form of disease. Influenza has visited the horse and proved either destructive, or has temporarily laid him aside for considerable periods of time; whilst rinderpest has kindly refused to ally itself to man or beast, save to the bovine species, and has shown at once how defined in its species and how diversified in the objects of selection are the subjects of epidemic disease.

Diseases of horses and cattle appear from time immemorial to have been closely allied to diseases in man, when either have assumed an epidemic form, since one has so frequently followed close upon the track of the other. Rome, the great nursery of epidemics in the form of dire pestilence, gave constant illustrations of this close connection between man and cattle. For brevity's sake, let one sample suffice:—" Annis 332, 296, and 291 B.C., Rome was again and again visited by pestilence, which was particularly fatal to *breeding women* and to *breeding cattle*. A similar visitation affected Rome, Anno 272 B.C."* Here *the identity of condition* in contrast with the distinction of zoological classification is so great, that we are almost led to the inference, which in later ages has demonstrated itself—that

---

* Bascome on "Epidemic Pestilence," 1851, page 10, to which work I am a great debtor.

diseases in distinct classes of animals may be so closely allied to each other that the lines of differentiation may culminate in the line of their mutual substitution, with the reduction of their mutual activities in the order of propagation; as in cow-pox, for small-pox, virus, by transplanting or exchange becoming nil for infection from the side of cattle to man, but analogous in differentiation or change in the nutritive functions of cell life, whereby resistance to repetition of the disease is obtained as completely in one form of the disease as the other, or as perfectly by the invader as the aboriginal, which demonstrates their proximity of parentage, or in familiar language, consanguinity.

The same author, Dr. Bascome, gives another illustration of the approximation in diseases of cattle to that of man, when it is present in the form of pestilence. He quotes from a Roman author, whose name is not given, but whose views are fully expressed in one short clause* :—" Pestilentia quæ priore anno ingruerat in boves, eo veteret in hominum morbos." However similar or dissimilar pestilence in cattle and man might be within the short space of one year, no doubt can exist, from the clause quoted, that the Roman writer considered that it was the same pestilence which was destructive to man that was the year previous destructive to horned cattle.

The general sentiment which is here conveyed by an ancient writer is fully substantiated by a long succession of epidemic periods since he wrote—namely, that epidemic disease in the bovine species has very frequently been preceded or followed by epidemic disease in man. In our own days this has been exemplified in the pleuro-pneumonia of cattle of 1846, preceding and following the cholera of 1849 and 1853, but since 1855 it has been gradually declining, but is more common in man of

---

* Dr. Bascome on " Epidemics," page 11, 1851.

late years, especially in its preliminary form of congestion of the lungs; whilst, on the other hand, the intense outbreak of small-pox in 1870, which but for vaccination would have been most fatal and extensive, was preceded by rinderpest in cattle, a disease the destructiveness of which was brought to a stand-still by that most perfect of all modes of isolation, burial—a method *scarcely practical* in its application to man. Just think how the world would be turned upside down if, to stamp out small-pox or scarlet fever, a common receptacle was prepared, and the infected were shot and buried, to check the spread of infection!!*

The possibility by isolation of stamping out infection in the *genus homo* is entirely impossible. The virulence may be mitigated, and in a measure limited; but, from civilization being a complete network of intercommunication in its most varied forms, the subtlety of infection must gain vent in some form or other. If it be limited to the chimney top, even there it would get a vent equal to a letter; and from thence infection would kindly shed its domestic favours upon some unsuspecting but highly susceptible object of its care and watchful solicitation.

But, whilst reflecting upon epidemics, it is interesting to observe how the principle of isolation applies itself to species, or a special class or order of life. Infection itself spreads from like to like, man to man, and cattle to cattle. Certain it is that such diseases as glanders pass from horses to man, and rinderpest from cattle to sheep, and hydrophobia from dogs and wolves to man; but its extension from the prime source to a second order or species of animal life checks its procreative properties, and so a limit in extension is effected.

Hence in all epidemics limitation of the vital force is

---

* See the several Reports upon the Origin and Nature of Cattle Plague in 1866, presented to both Houses of Parliament by command of Her Majesty.

isolated to given orders of organized beings and objects, and by fine forms of differentiation, amidst a thousand forms of vegetable growth, a potato or a wheat field is selected; and every spot of vegetable life surrounding this field, or particular class of vegetable, is growing in luxuriance, amidst the cold sheet of death which is shrouding in decay its smitten neighbour.

The fox runs to and fro from place to place, and, just infected by contact with a neighbouring fox, enters a hen roost, kills a hen, and leaves the rest, from fear of a surprisal by the farmer's dog, whose approach is known by his cry. Not a death occurs from his disease; no pheasants, hares, or ruminants of any kind fall victims to the distemper the fox holds within his active frame; but himself, and every fox he meets, falls a victim to that particular disease he has. Again, if it is not infectious, then all the more singular it is that an influence which is diffused over a large tract of land should only isolate one species; and, by some subtle differentiation in vital conditions between the fox and every living being by which the fox is surrounded, the evil or the destructiveness of the differentiation only falls to the special organism called the fox, whose shade of differentiation in vital power and function is bound by such fine lines of variation as to defy the most exact scan of human induction and human knowledge.

The tenants of the deep, from time to time, have shown the same tendency for epidemic influence to spend its devitalizing force, or to divert the vital force into an especial channel or order of organism. What applies to vegetables and animals applies with equal force to man. From one end of nosology to the other, in every widespread epidemic, we have the reiterated assurance that this form of disease is a part of the existing epidemic influence; and the other form of disease belongs to a distinct order of affection to the prevailing epidemic.

Hence, in extension of disease in epidemic periods, *isolation* is an essential feature in its diffusion and onward development. Moreover, it is common to have *excess of development in the insect world* and lower forms of animal life in particular kinds or grooves, *whilst a higher class, or man, falls a victim to special forms of disease in great numbers and wide areas;* but in such way that it is in its destructive powers subject to law and order. For what is a particular disease, but the imprint or law of that particular epidemic? And what is the diffusion of that disease but the order in which the law is extended and adjusted in its application?

Such a thing as a general diffusion of death or disease in equal ratio and proportion to all organized beings and objects, within a given area, is unknown and has no existence. Epidemium is always in its diffusion and application special and limited; and in its principle of distribution is conformable to the law of *isolation.*

Thus, in 1831-32 and 1849, cholera was a diversion of vital force in a particular form of limitation to human life; gangrene of the spleen in Russia; influenza in North Europe, 1837; potato famine in 1846 and rinderpest in 1869, are all samples of vital force acting in special forms of deterioration or perversion to health.

CONCERNING THE PROGRESS OF EPIDEMICS.

It is important to observe that upon the whole epidemics are progressive in their area of diffusion, and for the most part radiate from some particular or endemic area—as cholera from the Ganges, leprosy from Egypt, and the plague from the Levant. Also in their duration certain particular types last for centuries, then gradually decay.

Supposing these remarks are illustrated by four well-established diseases or epidemics—leprosy, Levant or bubonic plague, small-pox, and cholera—more light will be thrown upon the subject generally.

It will be found that, taking the year 1817 as a starting point for a new epidemic era, that *about* every 640 years there have been great epidemic changes or new forms of disease introduced, or old ones revived and spread over wider areas. Take the era from 1177 to 1817 as the last period. True plague shortly after this period or in 1217 visits Italy; 1222 Egypt, France, and Germany. In 1252 it visits England; 1347 this country is again visited, Italy again in 1477; but from 1517 till 1666 it was visiting all the great Continental towns, and from thence travelled into the surrounding countries, and from time to time paid a visit to this country, no country appearing to escape its awful calamities —Spain, Germany, France, the Netherlands, Italy, and many other lands, venting its last fury in this country in 1666. From this time it ceased in our land, but it returned to its native place in and about the Levant by way of Dresden, Dantzic, and Marseilles in 1720, Vienna and Hungary 1722, Moscow 1772, Egypt and Constantinople between 1778 and 1792; but since 1772 it has lingered much in Egypt and Asia Minor, where it still lingers, and often breaks out in a destructive manner in and about the locality of its first centre or endemic area.

Small-pox first appeared in England about the year 1174, or three years earlier than the fixed period, 1177. It was general throughout Europe in 1436, and in 1638 it got to America. If it came to England in 1174 it must have passed through France first, for it had been long before in Spain. Its widest diffusion and greatest mortality appear to have been attained in the seventeenth century, but at all times since its appearance north of the Apennines it has been a most fatal disease, breaking out in special parts in great severity until its arm of destruction was mutilated and impeded by vaccination. The years 1870 and 1871 would perhaps have seen as widespread destruction from this plague, as from any former visitation in Northern and Mid-Europe, but for vaccination.

There appears to be no authentic record of this disease passing further north than the Apennines and the Pyrenees from 1170 to 1174, which is not far off the hypothecated time of 1177, the assigned date or period for fresh epidemic outbreaks.

Leprosy is a chronic disease, and, as an endemic or epidemic disease, it never creates the same amount of fear and dread from its destructive powers, as diseases of more active and rapidly fatal tendencies. Its popularity, and the interest taken in it generally, arises more from the general abhorrence mankind have to all mutilations and diseases of the *skin;* there is no disease of the skin which creates more intense disgust, and about which infection is more entirely dreaded, than pertains to the cutaneous and constitutional disease called by the Greeks elephantiasis, and by the Latins lepra, or in modern times leprosy.

Its fatality does not appear to have been great at any period of history, and many of its victims linger on, an eye-sore to their friends and a burden to themselves. The leper, from being a nuisance to society, has become a trouble to the State, and Dr. Bascome informs us that so early as 1237 leprosy was a matter of legislature in England; and certain it is that immediately after the crusade under Richard the First, and known in history as the Third Crusade, leprosy was very common in this country and in France, which would be about the year 1190. It does not appear that the First Crusade of 1095 to 1099, under Godfrey de Bouillon, brought amongst the retainers in that mighty host, who returned to their native lands, the much-loathed disease, leprosy.

From the time of the Third Crusade to the end of the sixteenth century, leprosy was endemic in Great Britain, where there were many lazar-houses built for the reception of the sufferers from leprosy; but at the end of this period they gradually sank into the disreputable use of being a kind of casual-wards for vagrants, with old ulcers, and any loathsome self-created or natural ulcer on the skin, or cutaneous affection, the presence of which was supposed by the professional mendicants to afford an all-sufficient plea for why the opulent, out of their abundance, ought to give freely and ungrudgingly to the sufferers and distressed in their sore afflictions.

In 1477 leprosy appears to have been very prevalent in Spain; and in the city of Lebrija, in the province of Andalusia, so late as 1726 to 1764.

Abstracting 640 from 1177, leaves 537. The period, then, between 537 to 1177 next comes under review—a period of history the most replete with interest to the entire family of man, for it embraces the period of the planting and policy of the nations which rose upon the decay and downfall of the old Roman world, and introduces us to the infancy and childhood of modern Europe. But it also brings us back to the alphabet and spelling-books of *mental* infancy and *mental* childhood, and that to an extent which is almost incredible. Such historians as Hallam, Guizot, and Craik, with all their force of diction and illustration, almost fail to convince the manhood of Europe, that in childhood such imbecility and intellectual weakness could ever have been the lot of cultivated and refined modern Europe.

In, therefore, everything that is said upon this epoch on epidemics, such as choose to differ, especially after the eighth century, can do so with the utmost liberty, for variety of opinions are always admissible where data can scarcely be adduced.

Having so far prefaced the subject of the epoch from 537 to 1177, little else can be said beyond mere generalities.

It appears, then, that small-pox was first mentioned as occurring in Arabia 572; but how far it had spread, or where the first place of its appearance was noted, is not at all certain, but that it was recorded as existing in Arabia, 572 A.D.*

All history combines to fix its seat of origin in Arabia, and that this disease certainly commenced as an epidemic disease in the sixth century.

George Cedrenus, one of the Byzantine historians, who lived in the fifteenth century, a monk of a very legendary and speculative mind, mentions that the Emperor Diocletian died of small-pox; but when that monarch had laid down the imperial purple, and had retired to a private dwelling in the beautiful country of Salona, it must not be supposed he had no domestics, and that these domestics had no communication with the external world; that clothes were not washed; that tradespeople, retainers, and visitors never came to his residence; and, in the midst of such varied sources of infection, that the plague of small-pox was dropped from the clouds, and in a fine virgin soil it only touched one victim, and did not at the same time spread its baneful malignancy, like a devouring fire, on every side. Though Dr. Bascome mentions it as a matter of fact ("Opit. Cit.," page 22), and does not in the slightest refer to Diocletian's end being probably by his own hands, yet the authority and the entire circumstances of the case have such a legendary air about them, that the whole affair is not worthy of a minute's reflection.

But upon small-pox being an ancient disease, and antecedent to the sixth century, the decided judgment of Francis Adams, the learned editor and commentator upon

---

* Also see Paulus Ægineta, Vol. I., page 330, Sydenham Society's Edition.

Paulus Ægineta, is dead against it. He says:—" We may mention that, after having read, we may say, every word of every ancient writer on medicine that has come down to us, we can confidently affirm that the Greeks and Romans are altogether silent on the subject, and that we are indebted to the Arabians for the earliest accounts which we have of these diseases (small-pox and measles)."*

The earliest data we have of it is in a MS. at the University of Leyden, which gives the year 572, but how much earlier there are no certain means of determining. It was in Italy 614; and Spain 714, three years after the Saracens established themselves at Cordova, and conquered part of Spain. How it first reached Rome there is nothing of a decided character, but its being in Spain in 714 is directly traceable to Saracenic invasion. The conquerors brought the disease with them.

From these two dates, 614 and 714, till 1174, the historian has no certain grounds for speaking upon the plague of small-pox as having any regions beyond Italy and Spain, where this disease had spread itself.

The origin of the epidemic in Spain is apparent enough, because its spread would be coeval with its introduction by the Saracens, the same people amongst whom it first appeared in the sixth century. But it does not appear to have settled in France at that time; the southern mountain barriers of the Apennines, the Pyrenees, and the Alpine ranges gave small-pox its geographical boundary in Europe. But in the next period, from 1177 to 1817, it frequently scourged the entire of Europe, selecting at one time one country and then another, and at times the whole of Europe simultaneously, as in 1436.

So of the true Levant plague; it appeared between 543 and

---

* Paulus Ægineta, Vol. I., page 330, Sydenham Society's Edition.

566 in all countries bordering the Mediterranean, and southward, apparently by the way of Aden, on to India and Bombay. It visited Italy again in 614—if during this period it had ever entirely left it.

Though this disease broke out with such power and fatal effects in the sixth century, no traces of it appear to be found in Mid Europe, or to the North, till we reach the second period, between 1177 and 1817; during the first four centuries of which period it was a fatal scourge to every city in Europe, breaking out with unequal violence, and at unequal times, in this city or that country, and in this town or that village, sweeping the population off as with the besom of destruction.

Though negative evidence is the weakest of all evidence, yet the absence of all mention of Plague and Small-pox in the earlier historic records of Mid and Northern Europe is the only basis for supposing that they never reached these lands before 1177. It is still more to be regretted that history is more perfectly blank upon epidemics, which have slain their ten thousands, than about war, which has slain its one thousand, though the fears and customs of nations have received in their domestic habits more moulding and fashioning, and their destinies and successes have been more wrapped up in these fatal scourges, than by successive warlike contests; yet, with the exception of one or two historians, such as Thucydides and Procopius, the world, through history, would scarcely know that epidemics had ever existed, or they only existed for a few weeks or months, for the purpose of disappointing some prince, or duke, or ambitious robber from carrying his ill-conceived schemes and worse designs into immediate execution.

Leprosy is mentioned as known in Italy in 614, and in Spain lazar-houses were erected, 1067 A.D. But history is remarkably vacant in records of disease from 110 to 600, saving among the Saracens, several of whom give excellent descriptions of leprosy, and were familiar with it from Spain to Bagdad.

'As small-pox and plague, as described by the Arabian physicians and by Procopius, were unknown before this time, leprosy is the only disease through which we can trace the metamorphosis of disease in an old and well-authenticated malady, which sprang from the earliest land of wealth and civilization, and to this day remains, under every vicissitude of dynasty, the constant pest of its first endemic seat, the land of Egypt.

What further can be said upon epidemic eras must be chiefly confined, in the form of epidemic disease, to this old and despised disease, leprosy.

Strange to say, no veneration is paid to leprosy by the votaries of antiquity to this very day; though, as an old and somewhat transformed ailment, having for fashion's sake slowly moved with the times, it can claim respect and abhorrence from the Pharoahs on to the days of Queen Victoria.

To return, leprosy was in Spain 48 B.C., and was introduced there by Pompey's army. That army was in Asia Minor from 65 to 62 B.C. Pompey disbanded his army at Brundusium in 62 B.C., or in the same year that he left Asia Minor.

In this manner leprosy spread from Egypt to Italy and Spain, where it appears to have remained, or rather, as Tacitus informs us, in Italy it soon disappeared altogether; but in 614 it reappeared, and was so far prevalent that it received special consideration at the time. As this disease is slow in its spreading, and also, in addition to its admitted hereditary nature, is, as an endemic affection, *scarcely* conceded to be aided by infection, it may be presumed to have been slowly reviving for some 50 or 60 years earlier.

And the same may be said much later on with regard to Spain, for in 1067 lazar-houses were common in Spain; but they do not appear to have been restricted to the Arab population, and were asylums for the poor of all nationalities who suffered from this loathsome malady.

If infection be granted, France had been slow to receive the infection, as she was not under the necessity of resorting to lazar-houses till much later. If it was hereditary, then it had spread by intercourse very widely; and if endemic in Spain at this time, and for some time previously, the hereditary bias may be dispensed with, as it is hereditary in Norwegians now without any African admixture of blood, and its pure endemical origin excludes both Pompey and the Saracen from aiding it in any way whatever. But its hereditary nature would admit of a slow and very doubtful extension throughout Spain.

Upon the whole, Italy and Spain, both having had it within 60 B.C., and its reappearing after some 600 years of comparative absence, and being sufficiently frequent to claim for it a passing notice by the historian, and still more, a notice from existing rulers to build houses as asylums for those afflicted with it—all this shows that some change in those lands had occurred, no doubt of a subtle and obscure nature, which permitted and aided the spread of this awful chronic malady from the seventh century and onwards.

But let us enquire how Pompey's army became subject to this malady.

Egypt was the common focus or centre of leprosy, and so early as 1490 B.C., or earlier, it is admitted by the Egyptian historian Manetho to have an existence in his native land; but he fathers the disease upon the Hebrews, who brought it with them from Canaan.

Whether, therefore, the statement of Manetho is right or wrong in relation to the Hebrews, it is clear that Egypt had it from a very remote antiquity; and as Avicenna maintained it was endemic in Alexandria, and Lucretius stamps it as a disease originating within the influence of the Nile, it is, perhaps, not going too far to say that Egypt is its native soil, or its constant centre and endemic home.

It was evidently spread beyond its legitimate bounds in 26 A.D., because the New Testament refers to the cleansing of ten lepers in the land of Palestine about this period ; but what is much more to the point of time, is the spread of leprosy in Italy and Spain in the year 60 B.C. and onwards,* before which time Pliny affirms it was never known in Italy. He states it was imported from *Egypt* by Pompey the Great. Other historians have evidently followed Pliny in this matter, but it is a slight error.

Dr. Bascome ("Opit. Cit.," page 15) says that " the first appearance of leprosy in Spain coincides with its introduction into Italy, after having been prevalent in the army of Pompey the Great about sixty years, more or less, before the coming of Christ." Again, the same writer says :—" Anno 60 B.C., Spain, according to the opinion of several ancient and modern writers, both foreign and national, was one of the countries most subjected to the frightful disease of leprosy," where it has remained ever since.

Pompey was in Spain, as proconsul, from 76 to 71 B.C., and after defeating Perperna he returned with his victorious army to Italy. In 66 Pompey was appointed to terminate the war against Mithridates, which ended in 63 B.C. In 62 B.C. he returned to Italy, and at once disbanded his army at Brundusium. In 48 B.C. Pompey marched into Thessaly, at the head of an army of 40,000 (Plutarch), and was

---

* *Vide* Paulus Ægineta, Vol. II., page 6, Sydenham Society's Edition.

defeated in the neighbourhood of Pharsalia by Julius Cæsar; from thence he took ship for Egypt, and all but landed, when he was assassinated by the order of the Egyptian king. Pompey traversed Asia Minor, from Armenia and Pontus to Cœlo-Syria, Phœnicia, and Palestine. Twice during that time he wintered at Pontus; but at no time, not even in his death, did the sole of his *feet* rest in Egypt. Therefore, whatever disease Pompey's army brought to Italy, it did not contract it in Egypt (where Gabinius entered with a Roman army in 55 B.C.); but it was in Asia Minor where his victorious army came, and saw, and conquered, and contracted leprosy as a final legacy, the reward of living in an enemy's country.

It can be scarcely said that leprosy was endemic in Asia Minor, though God had threatened the Israelites with the diseases of Egypt, if they did not keep steadfast to his laws which he gave them by Moses. And we may infer that one of the curses inflicted upon his chosen people was the disease leprosy, which was contracted in Egypt.

Its spread in Asia Minor must have been very extensive in 66 B.C., and probably it had outstripped its old territory for some 40 years or so to have become so general that an army, well fed, clothed, and disciplined, should have become so affected by it that it became a focus of disease in other lands upon their return from active warfare.

Recruits from Nearer Spain and Italy composed the great bulk of Pompey's army, and on their being disbanded at Brundusium, whatever the more enterprising and active might do, the sick and disabled would seek for protection and quiet among their own relatives, and, according to the old and very popular notion, in their native air and country. In this manner the recruits from Spain who were lepers on their return would find their way back again to Spain, and those of Italy to their own relatives in their own country.

If it is contended that at a given period leprosy was endemic in Asia Minor, it must be said that its slow progress as an endemic disease, especially to strangers, is very marked. Take Bergen for a sample, where a stranger settling amongst them is scarcely ever known to become subject to it; much more a people whose habits, diet, active duties, and healthy or cheerful minds, elated by many successes, would be the last of all to be subjects for the slow induction of endemic disease of a very chronic character.

It could not be possibly of hereditary origin, because procreation with natives in four years could send back no men ready for soldiers, but little children, to Italy, old enough neither for training nor enlisting; and Pompey's army, when disbanded, would scarcely be burdened with little children. Pliny says, according to Adam's "Commentary" upon Paulus Ægineta, that when it was imported from Egypt it raged for a time, but soon became extinct.

By *raging* we understand that it spread itself to many persons in Italy, *irrespective of the soldiers;* or to large numbers beyond those who first brought the disease with them, who were of Pompey's army. Its rapid spread excludes hereditary origin, whilst its early extinction, or so far as not to be generally known, if at all, as at present in this country (England), indicates that its endemic form found in Italy an unsuitable soil where it might plant itself as a vigorous colony of blighted men. This view is confirmed by Celsus, who states that it "is a chronic disease, almost unknown in Italy, but very common in certain countries."

If, then, its origin was not endemic, nor yet hereditary, it must have been epidemic, or by some peculiar change in relation to the earth, whereby a disease, formerly settled within a given area, receives some fresh impulse and further powers of development than usual, and spreads, as it were, from land to land from a given centre, from some unknown agency, giving special activity, or power, to a poison when introduced into the system; which, in the case of leprosy, creeps over the whole frame, the kidneys excepted, and in time destroys its entire functions and vitality.

Not like cholera, and especially influenza—which, in a single night or a week, will attack a whole house, the side of a street, the wing of a barrack, or an apartment in a ship —some die, and some recover to propagate no evil to their offspring, unless cholera here and there *predisposes to malignant disease*, which, from several singular cases within personal knowledge, it almost appears to have this leaning at times; moreover, at times, adverse to the line of human intercourse, and at other times in obedience to it, if decaying animal matter is found in the same tract. But leprosy in 60 B.C. only followed the course of human intercourse, and that intercourse of a very definite and precise kind— veterans of a conquering army which laid at the feet of their commander the rich lands of the East.

Its sudden spread in Italy and its early decline *bespeak at once the infectious origin of a disease* in a land, at the time of its entrance, ungenial for its continuance as a fixed and local disease; whilst its appearance in Spain about the same time and from the same source, but abiding there, and also known in many lands (as Celsus informs us, and others later on), indicates that Spain (and North Africa, etc.) afforded for the new visitor the conditions required for a permanent and fruitful habitation.

Passing over a multitude of subsidiary matter, it may be said of leprosy that it is now very rarely infectious, but in its progress slow, and for the most part endemic; and in those parts where it is endemic its progress is mostly shown in its hereditary tendencies, but withdrawn from its endemic centres—as from Bergen, New Brunswick, etc.—and the offspring migrate to the Western States of America, or some genial locality for the sufferers, and its hereditary bias soon dies out. But in the East, in the South Sea Islands, in New Brunswick and Norway—in fact, in a range of wide circumference from its old centre the Nile—leprosy is showing itself as a widespread and inveterate disease, but at the present time rarely infectious.*

This disease, then, as the oldest-known disease, and whose centre is on the banks of the Nile, ought to show, if there is such a thing as metamorphosis of disease, indications of change of a slow and progressive character ought to exist ; and one of these changes appears to be that of infection, which it once possessed, but now has nearly lost. Moreover, its geographical distribution has assumed large proportions, and the victims it seizes are now becoming very numerous and alarming.

In the days of Moses, its external manifestations were considerably varied to leprosy as it now exists, or that described by the earlier Arabian, Roman, and Greek writers, and also of a very decidedly infectious character. That of Pompey's time was most likely a nearer approach to the Mosaic type than that which followed after, and was transitional in most of its outlines between the Mosaic and Greek or Arabian form, inasmuch as the disgust to the personal appearance is not so strongly marked as in earlier writers, nor yet the fear of infection ; yet the personal appearance to this day constitutes a very leading feature of disgust, and the avoidance of lepers, by later writers, is often referred to on this very ground.

It must be granted that the leprosy of Moses is only so

---

* The point of non-contagion in leprosy has been fairly disputed by Dr. G. A. Hansen, Bergen, Norway, in an able article " On the Etiology of Leprosy," in the *British and Foreign Medico-Chirurgical Review*, April, 1875.

far given as to describe *its first appearance*, and not its history or course in the least; since the object of the law relating to leprosy was to exclude all contact of lepers with healthy persons. It was one of those diseases of Egypt which the Israelites had to a limited degree in Egypt; and it is plain that its spread by infection ought and would be limited by isolation from the very first. And, secondly, that if a national calamity, there was no special virtue, if it was of Divine permission, why it should be increased by apathy and stulted moral appreciation of that which was chastisement from God, from that which was indifference to the rod and Him who used it.

Its fatal character is given us in the life of Uzziah, King of Judah, whose leprosy was directly of Divine infliction, who died a leper, and was buried apart in the burial-ground of the Kings of Judah (II. Chron. xxvi. 16—23). His sons being born antecedent to his disease, of course no hereditary taint could be transmitted. But the curse upon Gehazi, that "the leprosy therefore of Naaman shall cleave unto thee, and unto thy seed for ever"—*i.e.*, as long as his seed should continue to beget, so long should leprosy appear in his posterity, as for ever, here, means no more—implies an infliction more heavy than that which was usual to lepers; or, in other words, that in his family in particular it should be hereditary.

This also implies, what is now a mere matter of history, that leprosy does not destroy the power of procreation; it is quite possible it may limit it, since at its early advent victims of this disease are excessive in their sexual passions; hence one of its names, satyriasis. This peculiarity is given by Aretæus, and several Greek and Arabic writers.

But it is a remarkable fact that in the Levitical institutions not one word is said about the children of lepers, nor yet a word about the divorcing of lepers, or isolating their children, or forbidding them to marry; but if it was a disease so much excluded from ordinary intercourse with men that the Levitical law actually demanded that "he shall put a covering upon his upper lip, and shall cry, Unclean, unclean" (Leviticus xiii. 45), How is it that his children, if it was hereditary, were not included in the same condemnation?

But it will be said the fact of his being excluded the camp and dwelling alone, abhorred of men and women, would be sufficient to exclude marriage, or his wife remaining with him. This would answer very well if it were not certain that women occasionally have leprosy, and for them the same regulations pertained as were instituted for the male, and lepers of either sex were not forbidden to live together. Moreover, the Levitical law distinctly assumes that cases would present themselves of long standing, and which, through carelessness or indifference to the law, long escaped detection; for, Moses writes, " When the plague of leprosy is in a man, then he shall be brought unto the priest; and the priest shall see him: and, behold, if the rising be white in the skin, and it have turned the hair white, and there be quick raw flesh in the rising; it is an *old leprosy* in the skin of his flesh, and the priest shall pronounce him unclean, and shall not shut him up: for he is unclean" (Leviticus xiii. 9—11). The bare fact of no provision being made for hereditary leprosy, and the greatest caution being taken that it should not be approached within given limits by living cotemporaries, points precisely to the fact of its non-transmissibility, but to its present *infectious character*, probably from the breath as well as body exhalations; which view was entertained by Aretæus, Galen, and Avicenna, the latter mentioning that it is also endemic in Alexandria; and the Arabian, Alsaharavius, maintains three causes:—
1st, hereditary taint; 2nd, food; and 3rd, contagion

through the medium of respiration. The two last writers are some centuries later than either Aretæus or Galen, and wrote at a time when the Mosaic disease was passing from the infectious into one nearer to the modern form, which is truly endemic and hereditary, but rarely infectious.

This brief outline of the metamorphosis of the oldest disease in the world will give some idea of the constant changes which well-known diseases are prone to assume after the lapse of ages.

Modern pathologists, and may be nosologists, will maintain the invariability of disease; as some zoologists maintain the permanency of species in animals, not even admitting an improved breed by crossing, or the admission might be clothed in the language of a *forced deviation* from Nature by selection, contrary to the ordinary plan of Nature, but when left to itself would resolve itself back into the primary form of either one or the other of its progenitors.

But this is exactly the point aimed at, not to show that cholera ever becomes yellow fever, or that typhus fever is ever ague, though typhus fever may be nothing else than the epidemic and infectious form of endemic typhoid fever. The leading object designed is to show that vital power has very singular selective powers, shown alike in ordinary ailments, endemic and epidemic diseases; but that, with this great tendency to isolation in its destructive, or its preservative powers, yet that all diseases, in process of time, undergo modifications in type, duration, and activity, from constrained circumstances acting within, or upon the earth's surface; moreover, that changes pass over the whole globe, and affect every kind and form of disease, and may-be the entire of our vegetable economy in an almost imperceptible manner, ranging over extensive eras, and probably not far distant in duration from every 600 years; but, according to the nearest approximation, about every 640 years.

Again, in epidemic eras, generally towards the last 150 to 100 years, the dominant type begins to slowly decline, and its more active powers to abate; but it probably, after it first breaks out, does not get to its full vigour until 200 years from its start, and remains at that point for about 250 years, and then begins to decline. Every few years it spreads itself over wide areas of the earth's surface; and then again, as if ashamed to show its hydra head, like the serpularia, it draws in its widespread arms into the lands and places in which it is constantly settled as an endemic disease—the light of science, and may-be cries to Heaven for its abatement, both demanding its speedy subsidence.

Supposing it be correct to say that the lurking pest, leprosy, was roused from its Egyptian lair to spread by inroads upon the East and the West, and so included North Africa and Asia Minor in 103 B.C.; then in 60 B.C. the kindred soil to North Africa, *i.e.*, Spain, had it brought to it by its usual manner of extension—namely, infection; whilst in Italy it found an ungenial soil for its localization, as Pliny would make us believe. But from about 100 to 150 A.D. some indications are left of its having made itself well known in Italy, and means were put in practice for its amelioration; which, from the nature of things, is highly probable, considering at this time the comparatively settled state of the civilized world, and the greater facility in such case for the putting in practice sanitary measures.

Dr. Hecker has referred to the only one which was then adopted, which was judicious, and upon the whole comparatively efficient. He writes:—" Arrangements for the protection of the healthy against contagious diseases, the necessity of which is shown from these notions, were regarded by the ancients as useful; and by many, whose circumstances permitted it, were carried into effect in their houses. Even a total separation of the sick from the healthy, that indispensable means of protection against infection by contact, was proposed by physicians of the second century after Christ, in order to check the spreading of leprosy."

Of course, if it was getting less in the second century, the adoption of total separation of lepers from the healthy would not have been dreamed of. Neither would lazar-houses have been common in Spain in 1067 A.D. if for some long time before leprosy had not been on the increase in Spain, which, from 537 A.D., when history refers to its prevalence, would give a period of 530 years. It would be on the increase from about 537 to 600 A.D., and then more gradually; for it is some length of time before a chronic disease like leprosy makes much impression upon the community at large.

Again, we find England legislating for it in 1237, and Spain again in 1441. Of course in England it was new since 1177, and not till some centuries after its appearance in Spain was it to be found in England, though probably soldiers in the armies of Vespasian and Titus had remained some two to three years in Palestine before 70 A.D., who found their way back to Gaul and Britain as lepers at that early period, but died out without its spreading, from the ungenial condition of something in those lands at that time to its spread; but changes slowly going on within and on the earth's surface prepare lands for the reception and engrafting of disease, which at another time cannot be properly acclimatised. So we find that after 1177 an aptness for leprosy and other diseases had over-shadowed the variable clime and habits of the British people, and what the Crusaders got in Palestine was propagated and became indigenous to our land for some 300 years, after which time it slowly declined, and has now completely died out.

Of small-pox and Levant plague, we hear little, if anything, of them in North Europe from about 537 to 1174;

but 150 to 200 years after 1177—which is the next epidemic era —the already rising cloud begins to thicken up to 1670 to 1680, and thunderclap after thunderclap drops down its showers of pestilence and disease in the form of small-pox and plague in every city in Europe, and all the known world. Here in one ten years, there in another ten years, then twenty, then thirty, and then again simultaneously over all Europe and Asia; the most terrible and fatal storm of disease occurring in the form of Black death between 1348 and 1357, the three first years being the worst.

A curious and interesting enquiry presents itself in this Black death, which may be counted worthy of a passing notice, or, if it may be so styled, an attempt to break ice.

First: Can two diseases run parallel in the same body at the same time?

Second: Can two diseases amalgamate or coalesce, and out of that coalescence produce one new disease or hybrid?

First: Those who have seen much of scarlatina, measles, and small-pox, will have seen these affections running parallel with each other in the same person at the same time occasionally. Varicella, with measles or scarlatina, is not very infrequent; but the varicella in such cases is, though distinctive, of a very mild form, and the size of the pustule is very small. So far as personal knowledge is concerned, the *infection* is quite as active when combined as when either is distinct, but the severity of the attack is much milder and the duration shorter. The first few hours of appearance of the rash appear to be the most trying to the constitution; that passed, the two affections appear to modify each other, and to put a bar to each other's intensity; so that molecular change or cell nutrition undergoes less trial to its own processes of nutrition than when either runs its course alone. Variola and measles are rarely seen in the same person at the same time; and by the ancients, as they were cotemporaries in their time of appearance in Arabia, they were supposed (as Rhases and other Arabian physicians) to be different modifications of one general disease; but the duration of one or the other is so distinct that their mutual check upon each other is but for a short season, the variola occupying the same parts of the body when measles has disappeared as were occupied previously by that disease. Hence the check is merely a repression for a season, and does not appear to modify, hasten, or shorten the variola a single day.

When only one case of combined measles and small-pox has been seen it is not prudent to say much respecting it, but in giving an opinion, one thing only can be mentioned with certainty—namely, that the duration of the variola so far exceeding that of measles, it does not appear from its very nature to be so adapted to check and limit the other as diseases whose duration are much nearer to each other in running their course.

But in the Black death a very different tale must be told to that of measles and scarlatina together, or either of these with varicella.

Black death came apparently by way of China, according to Hecker, and thence by Persia to Asia Minor, and to Europe in 1348. Bascome gives it an African origin, and then to Asia; and Rapin, with his annotator, Tyndal, give it a Tartar origin, and from Cathay, in Asia, on to Constantinople and Europe. Its spread all over Asia, Europe, and Africa was more rapid and destructive than any pestilence before or since.

It was, in all respects, the Levant plague, first described by Procopius, save one, and that one was the lung affection—characterized by laborious respiration, much cough, and finally bloody expectoration, the sure sign of a fatal end, soon to be followed by hæmorrhage and death.

Gibbon, in his " Decline and Fall of the Roman Empire," copies directly from Procopius, who gives a very faithful outline of the plague as seen in his day. Let this description, then, suffice for our present purpose :—

" Æthiopia and Egypt have been stigmatized in every age, as the original source and seminary of the plague.* In a damp, hot, stagnating air, this African fever is generated from the putrefaction of animal substances, and especially from the swarms of locusts, not less destructive to mankind in their death than in their lives. The fatal disease which depopulated the earth in the time of Justinian and his successors,† first appeared in the neighbourhood of Pelusium, between the Serbonian bog and the eastern channel of the Nile. From thence, tracing as it were a double path, it spread to the east, over Syria, Persia, and

* I have read with pleasure Mead's short but elegant treatise, concerning " Pestilential Disorders," the eighth edition, London, 1722.

† The great plague which raged in 542, and the following years (Pagi, Critica, tom. ii, p. 518), must be traced in Procopius (Persic. lib. 2, c. 22, 23), Agathias (lib. 5, p. 153, 154), Evagrius (lib. 4, c. 29), Paul Diaconus (lib. 2, c. 4, p. 776, 777), Gregory of Tours (tom. ii., lib. 4, c. 5, p. 205), who styles it *Lues Inguinaria*, and the Chronicles of Victor Tununensis (p. 9, in Thesaur. Temporum), of Marcellinus (p. 54), and of Theophanes (p. 153). [The *Lacus Sirbonis* inspired terror among all the nations of antiquity. It was the fabled abode of Typhon, the evil genius of so many mythologies. Beneath its bed were boiling streams of bitumen and springs of raphtha, which often sent up lurid flames and heavy vapours; these were imagined to be the breath of the demon. (Herodotus, 2, 6; Plutarch, Anton., c. 3; Strabo. 16, 762.) In the course of ages this formidable lake was reduced within very narrow dimensions. (Pliny, 5, 14.) The retiring waters left a wide morass or bog, over which the winds spread the sands of the neighbouring desert, fatal to the unwary who ventured on their surface (Diodorus Siculus, 1, 30.) From this bog there issued, in the days of Justinian, a double miasma. The decaying exuviæ of the sea and the fumes of heated bitumen combined to impregnate the atmosphere with noxious vapours. These, inhaled by depressed and spirit-broken multitudes, living in filth, and indulging in the artificial excitement of stimulating drinks, produced the disease, no less by moral than by physical infection, which was carried, with such calamitous violence, from clime to clime. The ancient lake of Sirbonis has nearly, if not entirely, disappeared. (Cellarius, 2, 792.) But the name is still retained in maps, given to an apparently more recent collection of pools and lagunes, separated from the Mediterranean by a newly formed bank. These are called by the Turks, Sebâkhah Bardoual, or the lake of Baldwin, from that hero of the Crusades having died, when King of Jerusalem, in 1177, at the neighbouring town of Rhinocorura, the modern El Arisch. One of the latest and most authentic accounts of them may be found in the " Description de l'Egypte," drawn up from the official papers of the memorable French expedition (tom. xvi., p. 208).—Ed.

the Indies, and penetrated to the west, along the coast of Africa, and over the continent of Europe. In the spring of the second year, Constantinople, during three or four months, was visited by the pestilence; and Procopius, who observed its progress and symptoms with the eyes of a physician,* has emulated the skill and diligence of Thucydides in the description of the plague of Athens. The infection was sometimes announced by the visions of a distempered fancy, and the victim despaired as soon as he had heard the menace and felt the stroke of an invisible spectre. But the greater number, in their beds, in the streets, in their usual occupation, were surprised by a slight fever; so slight, indeed, that neither the pulse nor the colour of the patient gave any signs of the approaching danger. The same, the next, or the succeeding day, it was declared by the swelling of the glands, particularly those of the groin, of the arm-pits, and under the ear; and when these buboes or tumours were opened, they were found to contain a *coal*, or black substance, of the size of a lentil. If they came to a just swelling and suppuration, the patient was saved by this kind and natural discharge of the morbid humour. But if they continued hard and dry, a mortification quickly ensued, and the fifth day was commonly the term of his life. The fever was often accompanied with lethargy or delirium; the bodies of the sick were covered with black pustules or carbuncles, the symptoms of immediate death; and in the constitutions too feeble to produce an eruption, the vomiting of blood was followed by a mortification of the bowels. To pregnant women the plague was generally mortal; yet one infant was drawn alive from its dead mother, and three mothers survived the loss of their infected fœtus. Youth was the most perilous season, and the female sex was less susceptible than the male; but every rank and profession was attacked with indiscriminate rage, and many of those who escaped were deprived of the use of their speech, without being secure from a return of the disorder.† The physicians of Constantinople were zealous and skilful, but their art was baffled by the various symptoms and per-

* Dr. Freind (Hist. Medicin. in Opp., p. 416—420, Lond., 1733) is satisfied that Procopius must have studied physic, from his knowledge and use of the technical words. Yet many words that are now scientific were common and popular in the Greek idiom.

† Thucydides (c. 51) affirms that the infection could only be once taken; but Evagrius, born 536 A.D., who had family experience of the plague, observes, that some persons who had escaped the first, sank under the second attack; and this repetition is confirmed by Fabius.

tinacious vehemence of the disease; the same remedies were productive of contrary effects, and the event capriciously disappointed their prognostics of death or recovery. The order of funerals, and the right of sepulchres, were confounded; those who were left without friends or servants lay unburied in the streets, or in their desolate houses; and a magistrate was authorized to collect the promiscuous heaps of dead bodies, to transport them by land or water, and to inter them in deep pits beyond the precincts of the city. Their own danger, and the prospects of public distress, awakened some remorse in the minds of the most vicious of mankind; the confidence of health again revived their passions and habits; but philosophy must disdain the observation of Procopius, that the lives of such men were guarded by the peculiar favour of fortune or providence.*a* He forgot, or perhaps he secretly recollected, that the plague had touched the person of Justinian himself; but the abstemious diet of the emperor may suggest, as in the case of Socrates, a more rational and honourable cause for his recovery.* During his sickness the public consternation was expressed in the habits of the citizens; and their idleness and despondence occasioned a general scarcity in the capital of the East.

"Contagion is the inseparable symptom of the plague; which, by mutual respiration, is transfused from the infected persons to the lungs and stomach of those who approach them. While philosophers believe and tremble, it is singular that the existence of a real danger should have been denied by a people most prone to vain and imaginary terrors.† Yet the fellow-citizens of Procopius were satisfied,

---

Paullinus (p. 588). I observe that on this head physicians are divided; and the nature and operation of the disease may not always be similar.

* It was thus that Socrates had been saved by his temperance, in the plague of Athens. (Aul. Gellius, Noct. Attic., 2. 1.) Dr. Mead accounts for the peculiar salubrity of religious houses by the two advantages of seclusion and abstinence (p. 18, 19).

† Mead proves that the plague is contagious, from Thucydides, Lucretius, Aristotle, Galen, and common experience (p. 10—20); and he refutes (Preface, p. 2—13) the contrary opinion of the French physicians who visited Marseilles in the year 1720. Yet these were the recent and enlightened spectators of a plague which, in a few months, swept away fifty thousand inhabitants ("Sur la Peste de Marseille," Paris, 1786) of a city that, in the present hour of prosperity and trade, contains no more than ninety thousand souls. (Necker, "Sur les Finances," tom. i., p. 231.)

a It must be borne in mind in Gibbon's days that the assumed philosophers had put an extinguisher upon Providence, which has since been protected with a wet sheet, because the extinguisher was just beginning to melt from heat within.

by some short and partial experience, that the infection could not be gained by the closest conversation; and this persuasion might support the assiduity of friends or physicians in the care of the sick, whom inhuman prudence would have condemned to solitude and despair. But the fatal security, like the predestination of the Turks, must have aided the progress of the contagion; and those salutary precautions, to which Europe is indebted for her safety, were unknown to the government of Justinian. No restraints were imposed on the free and frequent intercourse of the Roman provinces; from Persia to France, the nations were mingled and infected by wars and emigrations; and the pestilential odour, which lurks for years in a bale of cotton, was imported, by the abuse of trade, into the most distant regions. The mode of its propagation is explained by the remark of Procopius himself, that it always spread from the sea-coast to the inland country; the most sequestered islands and mountains were successively visited; the places which had escaped the fury of its first passage, were alone exposed to the contagion of the ensuing year. The winds might diffuse that subtle venom; but, unless the atmosphere be previously disposed for its reception, the plague would soon expire in the cold or temperate climates of the earth. Such was the universal corruption of the air, that the pestilence, which burst forth in the fifteenth year of Justinian, was not checked or alleviated by any difference of the season. In time, its first malignity was abated and dispersed; the disease alternately languished and revived; but it was not till the end of a calamitous period of fifty-two years, that mankind recovered their health, or the air resumed its pure and salubrious quality. No facts have been preserved to sustain an account, or even a conjecture, of the numbers that perished in this extraordinary mortality. I only find that during three months, five, and at length ten, thousand persons died each day at Constantinople; that many cities of the East were left vacant, and that in several districts of Italy the harvest and the vintage withered on the ground. The triple scourge of war, pestilence, and famine, afflicted the subjects of Justinian; and his reign is disgraced by a visible decrease of the human species, which has never been repaired in some of the fairest countries of the globe."

Dr. Bascombe, in his "History of Epidemics," describes

the plague of 1348 as plague with a lung affection. "This malady," he says, "was accompanied by fever, difficulty of breathing, and spitting of blood; the respiration was so laborious that the sick were obliged to be always in an upright posture; deglutition was difficult, attended with flushed countenances and great restlessness; at the outset the cough was violent, but without loss of blood; after a short time, the expectoration becoming bloody, hæmorrhage succeeded, when death ensued in three days; spots and abscesses sometimes formed when the disease was protracted unto the fifth day." (Page 50.)

Though lengthened, an outline of the black death or plague of 1348 will be given from Dr. Hecker, translated by Dr. Babington in 1833, where a brief sketch is found of its history and nature, and its alliance to bubo and carbuncular plague, as first fully described by Procopius. Let it be observed that vomiting of blood, with mortification of the bowels, was occasionally present in the Levant or Justinian plague, but there is no mention of laboured breathing, spitting of blood, and then hæmorrhage, in the Justinian plague; but what hæmorrhage there was belonged rather to intense congestion in the veins of the stomach, ending in hæmatemesis and vomiting from impeded circulation in the liver and abdominal organs, but not at all arising from the lungs in any way whatever. Hence the lung affection in the Black death of 1348, and onwards, was an affection superadded to the Justinian plague.

"The nature of the first plague in China is unknown. We have no certain intelligence of the disease, until it entered the western countries of Asia. Here it showed itself as the Oriental plague, with inflammation of the lungs; in which form it probably also may have begun in China, that is to say, as a malady which spreads, more than any other, by contagion—a contagion that, in ordinary pestilences, requires immediate contact, and only under unfavourable circumstances of rare occurrence is communicated by the mere approach to the sick. The share which this cause had in the spreading of the plague over the whole earth was certainly very great; and the opinion that the Black death might have been excluded from Western Europe by good regulations, similar to those which are now in use, would have all the support of modern experience, provided it could be proved that this plague had been actually imported from the East, or that the Oriental plague in general, as often as it appears in Europe, always has its origin in Asia or Egypt. Such a proof, however, cannot be produced so as to enforce conviction; for it would involve the impossible assumption that either there is no essential difference in the degree of civilization of the European nations, in the most ancient and modern times, or that detrimental circumstances, which have yielded only to the civilization of human society and the regular cultivation of countries, could not formerly have maintained the bubo plague.

"The plague was, however, known in Europe before nations were united by the bonds of commerce and social intercourse;* hence there is ground for supposing that it sprung up spontaneously, in consequence of the rude manner of living and the uncultivated state of the earth—influences which peculiarly favour the origin of severe diseases. Now, we need not go back to the earlier centuries, for the 14th itself, before it was half expired, was visited by five or six pestilences.†

"If, therefore, we consider the peculiar property of the plague that, in countries which it has once visited, it remains for a long time in a milder form, and that the epidemic influences of 1342, when it had appeared for the last time, were particularly favourable to its unperceived continuance till 1348, we come to the notion that in this eventful year,

---

* According to Papon, its origin is quite lost in the obscurity of remote ages; and even before the Christian era we are able to trace many references to former pestilences. "De la peste, ou époques mémorables de ce fléau, et les moyens de s'en préserver." T. II., Paris, An. VIII. de la rép. 8.

† 1301, in the south of France; 1311, in Italy; 1316, in Italy, Burgundy, and Northern Europe; 1335, the locust years, in the middle of Europe; 1340, in Upper Italy; 1342, in France; and 1347, in Marseilles and most of the larger islands of the Mediterranean. Ibid., T. II., p. 273.

also, the germs of plague existed in Southern Europe, which might be vivified by atmospherical deteriorations; and that thus, at least in part, the Black plague may have originated in Europe itself. The corruption of the atmosphere came from the East; but the disease itself came not upon the wings of the wind, but was only excited and increased by the atmosphere where it had previously existed.

"This source of the Black plague was not, however, the only one; for, far more powerful than the excitement of the latent elements of the plague by atmospheric influences was the effect of the contagion communicated from one people to another on the great roads, and in the harbours of the Mediterranean. From China the route of the caravans lay to the north of the Caspian Sea, through Central Asia, to Tauris. Here ships were ready to take the produce of the East to Constantinople, the capital of commerce, and the medium of connection between Asia, Europe, and Africa.* Other caravans went from India to Asia Minor, and touched at the cities south of the Caspian Sea, and lastly, from Bagdad, through Arabia to Egypt; also the maritime communication on the Red Sea, from India to Arabia and Egypt, was not inconsiderable. In all these directions contagion made its way; and doubtless Constantinople and the harbours of Asia Minor are to be regarded as the foci of infection, whence it radiated to the more distant seaports and islands.

"To Constantinople, the plague had been brought from the northern coast of the Black Sea,† after it had depopulated the countries between those routes of commerce; and appeared, as early as 1347, in Cyprus, Sicily, Marseilles and some of the seaports of Italy. The remaining islands of the Mediterranean, particularly Sardinia, Corsica, and Majorca, were visited in succession. Foci of contagion existed also in full activity along the whole southern coast of Europe; when, in January, 1348, the plague appeared in Avignon,‡ and in other cities in the south of France and north of Italy, as well as in Spain.

"The precise days of its eruption in the individual towns are no longer to be ascertained, but it was not simultaneous; for in Florence, the disease appeared in the beginning of

---

* Compare Deguignes, Loc. cit., p. 288.
† According to the general Byzantine designation, "from the country of the hyperborean Scythians." Kantakuzen, Loc. cit.
‡ Guid. Cauliac, Loc. cit.

April;* in Cesena, the 1st of June;† and place after place was attacked throughout the whole year, so that the plague, after it had passed through the whole of France and Germany, where, however, it did not make its ravages until the following year, did not break out till August in England; where it advanced so gradually, that a period of three months elapsed before it reached London.‡ The Northern Kingdoms were attacked by it in 1349—Sweden, indeed, not until November of that year; almost two years after its eruption in Avignon.§ Poland received the plague in 1349, probably from Germany,‖ if not from the northern countries; but in Russia, it did not make its appearance until 1351, more than three years after it had broken out in Constantinople. Instead of advancing in a north-westerly direction from Tauris and from the Caspian Sea, it had thus made the great circuit of the Black Sea, by way of Constantinople, Southern and Central Europe, England, the Northern Kingdoms and Poland, before it reached the Russian territories; a phenomenon which has not again occurred with respect to more recent pestilences originating in Asia.

"Whether any difference existed between the indigenous plague, excited by the influence of the atmosphere, and that which was imported by contagion, can no longer be ascertained from the facts; for the contemporaries, who in general were not competent to make accurate researches of this kind, have left no data on the subject. A milder and a more malignant form certainly existed, and the former was not always derived from the latter, as is to be supposed from this circumstance—that the spitting of blood, the infallible diagnostic of the latter, on the first breaking out of the plague, is not similarly mentioned in all the reports; and it is therefore probable, that the milder form belonged to the native plague—the more malignant, to that introduced by contagion. Contagion was, however, in itself, only one of many causes which gave rise to the Black plague.

"This disease was a consequence of violent commotions in the earth's organism—if any disease of cosmical origin can be so considered. One spring set a thousand others in

---

* Matt. Villani, Istorie, in Muratori, T. XIV., p. 14.
† Annal. Caesenat., Ibid., p. 1179.
‡ Barnes, Loc. cit.
§ Olof Dalin's, " Svea-Rikes Historie," III. vol., Stockholm, 1747—61, 4. Vol. II., C. 12, p. 496.
‖ Dlugoss, " Histor. Polon.," L. IX., p. 1086, T. I. Lips., 1711, fol.

motion for the annihilation of living beings, transient or permanent, of mediate or immediate effect. The most powerful of all was contagion; for in the most distant countries, which had scarcely yet heard the echo of the first concussion, the people fell a sacrifice to organic poison—the untimely offspring of vital energies thrown into violent commotion."

Comparing the plague of Justinian of 526 A.D., which is Gibbon's date for that plague, with the Black death of 1348, it may be justly stated that both were bubonic and carbuncular plagues, with a superadded lung disease in the latter epidemic, which was as diagnostic in its laboured breathing and spitting of blood as the presence of buboes, etc., were in true Levant plague; but carbuncles appear to have been less prevalent in the latter epidemic than in the Justinian one.

The general conclusion arrived at by comparing these two well-authenticated and carefully described epidemics is this, that distinct and specific affections in themselves occasionally unite to constitute one continuous disease, by which the entire system is pervaded and brought under complete sway and conquest; and in this new form of infection and widespread diffusion, at least for a season or considerable duration of time, as of several years, can be induced and sustained.

A modification by blending appears to be the natural result of coalescence; hence in the majority of cases, even where convalescence was established, carbuncles never appeared, though buboes in the neck, armpits, and inguinal regions were common.

Having advanced the hypothesis of hybrid diseases, as distinct from parallel diseases running their course, though modifying one another in one and the same person at the same time, it may be asked, Is there any trace of hybridity in any other disease than the Black plague, which probably was the blending of an infectious lung disease of a more or less endemic character, common in the extensive plains of Tartary, with another infectious disease known as the Levant or Justinian plague; the specific poisons at some point of junction blending their primitive blastema, or something equally incomprehensible (beyond the notion of their having opposite centres of attraction, which will be equal to mutual coalescence), and when blended being sustained in their integrity by an exact epidemic constitution of atmosphere adapted for their growth and diffusion, till every nook and corner is reached where man holds intercourse with his brother man? Let this be briefly considered.

The Black death in all lasted about fifty years; its greatest destructive power was in the first few years of its advent into Europe. About a hundred years after its disappearance, the old plague took its usual course of visiting one city for a year or two, another in a few years after, then diffusing itself over a particular country.

But we find a new disease showing itself about the year 1494, though Florence and France appear to have shown indications of this disease towards the end of the twelfth century; and one of the Medici has honoured posterity by having his name enrolled among the noble patricians of Italy who fell a victim to *Syphilis* before it was generally known as the *mal de France*.

Was this new disease a blending of plague and leprosy in one new entity, or a disease retaining in itself some faint outlines of both in a modified form?

There is no disease in which contagion or direct local contact is more essential to be subject to its infectious properties than is syphilis. Ancient writers upon leprosy appear to have judged infection arose from the breath and body exhalations of the sufferer, and so of plague. Doubts appear to be entertained whether now either of these diseases are infectious—above all, leprosy. Plague is not hereditary, but leprosy is decidedly so. One disease is acute, and the other chronic. Both appear to be blood diseases, but leprosy, according to the ancients, was an universal and internal cancer, working from within outwards, and laying hold of every vital structure, which moderns have confirmed save with regard to the kidneys. Plague was a blood and lymphatic disease; therefore its poison was prone to reach certain glands, as of the inguinal and axillary regions and the neck.

Syphilis, in its secondary or tertiary forms, is a blood disease. Modern pathology has detected syphiloid infiltration, or morbific deposits, in almost every structure of the body; it is of limited hereditary transmission, and whether it ever reaches the second generation is extremely doubtful, though testimony here and there might lead to such an inference; but careful observation, and a little ordinary discernment, might lead to a wholesome doubt as to the value of the testimony given.

All matter from an exposed surface upon the extremities is prone to excite inflammation of lymphatic glands in the inguinal or axillary regions; but the syphilitic poison is perhaps more prone than that of any other kind to excite inflammation in the glands, and develop ordinary buboes. Occasionally the soft superficial chancre will produce true secondary symptoms without buboes appearing, and will infect another person. Mr. A. had a soft chancre on the penis at its middle; its greatest width was transversely to its long axis, never was hard, healed perfectly, and was followed by eruption and ulcerated sore throat, and thinning of the hair, all of which followed within three months from the time the chancre healed. Mrs. A. followed in the train, but a month later; but got more distinct bald patches on the head, and the hair very scanty elsewhere. Fifteen months after, a son was born, and in six weeks a wretched syphilitic eruption appeared on the nates and arms, which was long in being subdued. Here is a distinct case of secondary syphilis from soft chancre without bubo, and it is not the only one observed; but in plague, occasionally, buboes will not present themselves, yet the nature of the affection, and its fatal end, are quite as certain as when bubo has appeared.

If it be granted that syphilis is a hybrid disease of plague and leprosy, it must be added that their amalgamation has modified and limited both alike in relation to fatality and permanency of duration; and in this respect it has much the characteristics of hybridity in a general way. The duration of transmission from generation to generation is materially abridged, and the extremes of development are contracted and modified; different to that of the Black death, in which, though the persistency of the disease was not so great in duration as the common plague, yet its rapidity of extension, contrary to hybridity generally, was increased rather than diminished; but this may be explained by both poisons in their essential constitutions being but mere varieties of a common species, both being essentially blood poisons of an acute character, and in all essential points of propagation and alimentation (if such a word for a poison increasing by feeding may be used) closely allied, and, as before said, mere varieties of the same species; then, as under favourable conditions of cross-breeding, increase and multiplication would go on with redoubled energy to that which either, in an isolated condition, could accomplish.

It is pleasing to observe, in the view here expressed of the origin of syphilis, a somewhat similar notion is given by Francis Adams in his "Annotations to Paulus Ægincta." He writes:—" By the way, we may be permitted to state that we have long been convinced that the syphilis of modern times is a modified form of ancient elephantiasis (leprosy of Arabs)." This opinion is maintained by several of the writers of the Aphrodisiacus, and also by the learned Sprengel, who gives a very interesting disquisition on syphilis in his " History of Medicine."

The opinion already given about syphilis being a hybrid between leprosy and plague, arose from the reading of history generally, and appeared to be forced upon the attention by a close analysis of a large number of dissimilar and closely co-related facts, scattered over a great variety of historical incidents, from the time of Constantine the Great on to our own times; and it was pleasant to find the same foot-path had been trodden by men whose learning and judgment are deserving of *much* respect, and whose sentiments were hailed as a confirmation that both were going at least part way on the same journey. There is yet another disease whose origin is interesting and singular, and for the details of which we have an eye-witness and personal sufferer, whose powers of observation and carefulness of description place him at the head of all chroniclers of disease that history has given. This disease is the Athenian plague, as recorded by Thucydides.*

"When they had not been many days in Attica, the plague first began to show itself amongst the Athenians; though it was said to have previously lighted on many places about Lemnos and elsewhere. Such a pestilence, however, and loss of life as this was nowhere remembered to have happened. For neither were physicians of any avail at first, treating it as they did, in ignorance of its nature—nay, they themselves died most of all, inasmuch as they most visited the sick—nor any other art of man. And as to the supplications that they offered in their temples, or the divinations, and similar means, that they had recourse to, they were all unavailing; and at last they ceased from them, being overcome by the pressure of the calamity.

---

* After comparing many translations, epitomes, and compliments to Lucretius, the introduction of a verbal and exact translation of Thucydides' description of the plague of Athens, by the Rev. H. Dale, is perhaps the fairest way of examining the whole subject.

"It is said to have first begun in the part of Æthiopia above Egypt, and then to have come down into Egypt, and Libya, and the greatest part of the king's territory. On the city of Athens it fell suddenly, and first attacked the men in the Piræus; so that it was even reported by them that the Peloponnesians had thrown poison into the cisterns; for as yet there were no fountains there. Afterwards it reached the upper city also; and then they died much more generally. Now let every one, whether physician or unprofessional man, speak on the subject according to his views; from what source it was likely to have arisen, and the causes which he thinks were sufficient to have produced so great a change [from health to universal sickness]. I, however, shall only describe what was its character; and explain those symptoms by reference to which one might best be enabled to recognise it through this previous acquaintance, if it should ever break out again; for I was both attacked by it myself, and had personal observation of others who were suffering with it.

"That year then, as was generally allowed, happened to be of all years the most free from disease, so far as regards other disorders; and if any one *had* any previous sickness, all terminated in this. Others, without any ostensible cause, but suddenly, while in the enjoyment of health, were seized at first with violent heats in the head, and redness and inflammation of the eyes; and the internal parts, both the throat and the tongue, immediately assumed a bloody tinge, and emitted an unnatural and fetid breath. Next after these symptoms, sneezing and hoarseness came on; and in a short time the pain descended to the chest, with a violent cough. When it settled in the stomach, it caused vomiting; and all the discharges of bile that have been mentioned by physicians succeeded, and those accompanied with great suffering. An ineffectual retching also followed in most cases, producing a violent spasm, which in some cases ceased soon afterwards, in others much later. Externally the body was not very hot to the touch, nor was it pale; but reddish, livid, and broken out in small pimples and sores. But the internal parts were burnt to such a degree that they could not bear clothing or linen of the very lightest kind to be laid upon them, nor to be anything else but stark naked; but would most gladly have thrown themselves into cold water if they could. Indeed, many of those who were not taken care of did so, plunging into cisterns in the agony of their unquenchable thirst; and it was all the same whether they drank much or little. Moreover, the misery of restlessness and wakefulness continually oppressed them. The body did not waste away so long as the disease was at its height, but resisted it beyond all expectation: so that they either died in most cases on the ninth or the seventh day, through the internal burning, while they had still some degree of strength; or if they escaped [that stage of the disorder], then, after it had further descended into the bowels, and violent ulceration was produced in them, and intense diarrhœa had come on, the greater part were afterwards carried off through the weakness occasioned by it. For the disease, which was originally seated in the head, beginning from above, passed throughout the whole body; and if any one survived its most fatal consequences, yet it marked him by laying hold of his extremities; for it settled on the pudenda, and fingers, and toes, and many escaped with the loss of these, while some also lost their eyes. Others, again, were seized on their first recovery with forgetfulness of everything alike, and did not know either themselves or their friends.

"For the character of the disorder surpassed description; and while in other respects also it attacked every one in a degree more grievous than human nature could endure, in the following way especially, it proved itself to be something different from any of the diseases familiar to man. All the birds and beasts that prey on human bodies, either did not come near them, though there were many lying unburied, or died after they had tasted them. As a proof of this, there was a marked disappearance of birds of this kind, and they were not seen either engaged in this way, or in any other; while the dogs, from their domestic habits, more clearly afforded opportunity of marking the result I have mentioned.

"The disease, then, to pass over many various points of peculiarity, as it happened to be different in one case from another, was in its general nature such as I have described. And no other of those to which they were accustomed afflicted them besides this at that time; or whatever there was, it ended in this. And [of those who were seized by it] some died in neglect, others in the midst of every attention. And there was no one settled remedy, so to speak, by applying which they were to give them relief; for what did good to one, did harm to another. And no constitution showed itself fortified against it, in point either of strength or weakness; but it seized on all alike, even those that were treated with all possible regard to diet. But the most dreadful part of the whole calamity was the dejection felt whenever any one found himself sickening (for by immediately falling into a feeling of despair, they abandoned themselves much more certainly to the disease, and did not resist it), and the fact of their being charged with infection from attending on one another, and so dying like sheep. And it was this that caused the greatest mortality amongst them; for if through fear they were unwilling to visit each other, they perished from being deserted, and many houses were emptied for want of some one to attend to the sufferers; or if they did visit them, they met their death, and especially such as made any pretensions to goodness; for through a feeling of shame they were unsparing of themselves, in going into their friends' houses [when deserted by all others]; since even the members of the family were at length worn out by the very moanings of the dying,* and were overcome by their excessive misery. Still more, however, than even these, did such as had escaped the disorder show pity for the dying and the suffering, both from their previous knowledge of what it was, and from their being now in no fear of it themselves; for it never seized the same person twice, so as to prove actually fatal. And such persons were felicitated by others; and themselves, in the excess of their present joy, entertained for the future also, to a certain degree, a vain hope that they would never now be carried off even by any other disease.

" In addition to the original calamity, what oppressed them still more was the crowding into the city from the country, especially the new comers. For as they had no houses, but lived in stifling cabins at the hot season of the year, the mortality amongst them spread without restraint; bodies lying on one another in the death-agony, and half-dead creatures rolling about in the streets and round all the fountains, in their longing for water. The sacred places also in which they had quartered themselves, were full of the corpses of those that died there in them; for in the surpassing violence of the calamity, men, not knowing what was to become of them, came to disregard everything, both sacred and profane, alike. And all the laws were violated which they before observed respecting burials; and they buried them as each one

---

* Or, "by lamenting for the dying." See Arnold's note.

could. And many, from want of proper means, in consequence of so many of their friends having already died, had recourse to shameless modes of sepulture; for on the piles prepared for others, some, anticipating those who had raised them, would lay their own dead relative and set fire to them; and others, while the body of a stranger was burning, would throw on the top of it the one they were carrying, and go away.

"In other respects also the plague was the origin of lawless conduct in the city, to a greater extent [than it had before existed]. For deeds which formerly men hid from view, so as not to do them just as they pleased, they now more readily ventured on; since they saw the change so sudden in the case of those who were prosperous and quickly perished, and of those who before had had nothing, and at once came into possession of the property of the dead. So they resolved to take their enjoyment quickly, and with a sole view to gratification; regarding their lives and their riches alike as things of a day. As for taking trouble about what was thought honourable, no one was forward to do it; deeming it uncertain whether, before he had attained to it, he would not be cut off; but everything that was immediately pleasant, and that which was conducive to it by any means whatever, this was laid down to be both honourable and expedient. And fear of gods, or law of men, there was none to stop them; for with regard to the former they esteemed it all the same whether they worshipped them or not, from seeing all alike perishing; and with regard to their offences [against the latter], no one expected to live till judgment should be passed on him, and so to pay the penalty of them; but they thought a far heavier sentence was impending in that which had already been passed upon them; and that before it fell on them, it was right to have some enjoyment of life.

"Such was the calamity which the Athenians had met with, and by which they were afflicted, their men dying within the city, and their land being wasted without. In their misery they remembered this verse amongst other things, as was natural they should; the old men saying that it had been uttered long ago—

"'A Dorian war shall come, and plague with it.'"

Now there was a dispute amongst them [and some asserted] that it was not a 'plague' [*loimos*] that had been mentioned in the verse by the men of former times, but 'a famine,' *limos*]; the opinion, however, at the present time naturally prevailed that 'a plague' had been mentioned, for men adapted their recollections to what they were suffering. But, I suppose, in case of another Dorian war ever befalling them after this, and a famine happening to exist, in all probability they will recite the verse accordingly. Those who were acquainted with it recollected also the oracle given to the Lacedæmonians, when on their inquiring of the god whether they should go to war, he answered, 'that if they carried it on with all their might, they would gain the victory, and that he would himself take part with them in it.' With regard to the oracle, then, they supposed that what was happening answered to it. For the disease had begun immediately after the Lacedæmonians had made their incursion; and it did not go into the Peloponnese, worth even speaking of, but ravaged Athens most of all, and next to it the most populous of the other towns. Such were the circumstances that occurred in connection with the plague.

"The Peloponnesians, after ravaging the plain, passed into the Paralian territory, as it is called, as far as Laurium, where the gold mines of the Athenians are situated. And first they ravaged the side which looks towards Peloponnese; afterwards, that which lies towards Euboea and Andrus. Now, Pericles being general at that time as well as before, maintained the same opinion as he had in the former invasion, about the Athenians not marching out against them.

"While they were still in the plain, before they went to the Paralian territory, he was preparing an armament of a hundred ships to sail against the Peloponnese; and when all was ready, he put out to sea. On board the ships he took four thousand heavy-armed of the Athenians, and three hundred cavalry in horse-transports, then for the first time made out of old vessels; a Chian and Lesbian force also joined the expedition with fifty ships. When this armament of the Athenians put out to sea, they left the Peloponnesians in the Paralian territory of Attica. On arriving at Epidaurus, in the Peloponnese, they ravaged the greater part of the land, and having made an assault on the city, entertained some hope of taking it; but did not, however, succeed. After sailing from Epidaurus, they ravaged the land belonging to Trœzen, Haliæ, and Hermione; all which places are on the coast of the Peloponnese. Proceeding thence they came to Prasiæ, a maritime town of Laconia, and ravaged some of the land, and took the town itself, and sacked it. After performing these achievements, they returned home; and found the Peloponnesians no longer in Attica, but returned.

"Now all the time that the Peloponnesians were in the Athenian territory, and the Athenians were engaged in the expedition on board their ships, the plague was carrying them off both in the armament and in the city, so that it was even said that the Peloponnesians, for fear of the disorder, when they heard from the deserters that it was in the city, and also perceived them performing the funeral rites, retired the quicker from the country. Yet in this invasion they stayed the longest time, and ravaged the whole country; for they were about forty days in the Athenian territory."

The peculiarities of this plague may be put forward in the following manner. It began at the head and eyes, which latter were red and inflamed; the tongue and throat were tinged with bloody exudation and fetid breath. It then went to the throat and chest, occasioning hoarseness and violent coughing. Next it came to the stomach, occasioning vomiting. At this stage, which was the seventh to ninth day, many died; but if he survived, the malady descended lower, when diarrhœa came on, and probably caused, as Thucydides affirms, ulceration of the bowels. Finally, many of those who survived these sufferings had not yet seen the end of the disease, for the poison of this plague, whatever it might be, destroyed the integrity of structures remote from the head, the first seat of its attack; and in the fingers and toes, as well as the generative organs, amputation by sphacelus was not infrequent; and occasional destruction of the eyes, and not merely of the sight, completed the sad wreck which befel the survivors of this awful malady—awful alike in its destruction of life, and awful, at times, in the wrecks of men who survived its invasion.

To this category of evils, which in succession laid hold upon the sufferers, let it be noticed that "externally the body was not very hot to the touch, nor was it pale; but reddish, livid, and broken out in small pimples and sores." Adams calls the pimples and sores phlyctæna and ulcers:—"The skin reddish or livid, and covered with minute phlyctæna and ulcers." (Paulus Agincta, Vol. II., page 279, Sydenham Society's Edition.)

Here is a disease of which its counterpart cannot be found in modern times. In one point it is evidently near to small-pox—the livid skin from laboured respiration, arising chiefly from the state of the throat, with the small pimples and sores, in different stages of ripeness; the marked delirium which mostly accompanies this disease; its order of progress from the head to the feet, as is the case in most eruptive diseases of an infectious character, as measles, scarlatina, and small-pox, etc.; vomiting and bowel affections being occasionally very marked, and trying symptoms, or rather conditions of the disease; and, as a finale, where recovery takes place, the occasional loss of sight. On the other hand, mortification of the fingers, toes, and pudenda no more belong to small-pox than does phlegmasia dolens to typhus fever or croup.

Again, the duration of the disease is too short for variola, when it proves fatal.

What can the true etiology of this disease be? Conjecture can alone be given, but no positive dictum, unless it be in the highest degree of a presumptive and ignorant nature.

Suffer fools to speak, as it gives the wise an opportunity of correcting and setting matters straight, after they have been all disordered by consummate folly.

For years an idea has presented itself that diseases occasionally converge, as in syphilis, as here given, and that occasionally they diverge or fork outwards from a given centre.

Rhases and Avicenna, and all the Arabian physicians, viewed measles and small-pox as having a common origin in different conditions of the bile, but both had their origin in the bile; and small-pox and measles were always considered as two distinct, but closely allied, or twin diseases. That both affect the throat, and chest, and the eyes is certain, and also the skin. But the manner and degree of affecting those parts, in many points, are widely different; yet in the mode of termination, when fatal, very similar in some points. In both recession of the eruption is most dangerous; and in both the breathing is often the chief and most important indication of a serious or fatal termination. These twin diseases appeared in Arabia almost, if not quite simultaneously; both were skin diseases in a very important respect, and both were infectious, and as a rule going about, or were epidemic at the same time. In short, the inference was drawn that they had one common origin, and that their origin *was in an older and distinct disease*, which in lapse of ages had varied considerably; and finally, at a new epidemic era, had resolved itself into two free and independent diseases, now known as measles and small-pox.

The *livid reddish skin*, with pimples and sores, indicates that two forms of eruption ran parallel in the same cases— the pimples and the reddish skin; and from the minuteness of the observer it is scarcely admissible that in the short space of seven days the fact of one being the forerunner of the other, and only different in degrees of age, could have escaped observation and careful discrimination; but though the disease presented two forms of eruption, yet the distinctive character of the disease was that it should have this specific form of dissimilar sores, *or spots, and reddish skin*, and the very absence of it ought only to indicate some modification of the disease in *particular individuals*, in whom the admixture of kinds of spots or sores were not observable when under the influence of the plague at that time.

So much for the eruption; but neither small-pox nor measles after recovery are followed by mortification of the fingers and toes, etc. Neither, as a rule, is small-pox in itself so short in its course as the seventh to ninth day, unless something very unusual should occur to cut it short.

Ergotism is a disease in which this sphacelus can occur as the result of diseased rye eaten as common food. This has occurred several times at Sologne and other parts of France; and in our own country this disease occurred in the family of John Downing, of Watlesham, so minutely described by Wollaston. The rapidity with which sphacelus occurred, after first being seized with pain, appears to have been most marked, it happening so early as the fifth and sixth day. But ergotism is neither infectious, nor yet attended with any high fever, and it is entirely endemic. But the plague of Athens was an infectious disease, and had spread from Ethiopia to Greece, and who knows how much further, and was a disease attended with delirium, intense thirst, and very oppressed breathing; hence the bringing into play as a cause a special diet will in nowise account for this extraordinary phenomenon of mortification of the extremities.

If it was not in all respects a disease *sui generis*, we must supply for it a source which had both an infectious nature and also made the extremities and remote parts its special object of attack; and if we are to say there was at that time a disease which possessed that property in a special manner, that disease must be admitted to be modified and accelerated infection of leprosy poison, blending as a hybrid, with a rubeoloid affection of an infectious character also, and in many respects simulating, in the kind of its eruption, variola.

The Athenian plague has no counterpart in modern times, and cannot be compared to any one single existing disease.

That it should now be extinct is no argument against its former existence, if that plague was truly an animal poison of the hybrid character, and of an infectious nature; for either the breed will go on till one or other lapses into the primitive species, and so continues for its natural period, as the Black death returning to simple Levant plague, or the disease has but a limited duration, and, by the reproductive power in hybridity being weaker than in the pure, extinction follows as a sequence and a check to vigorous reproduction.*

Upon what grounds, it will be asked, is a chronic affection made to ally itself to an acute disease, and to run a course of such fearful haste as to do more in one month than in its pure form it does in ten, twenty, or more years? Its chronic spongy gums, and its chronic bowel affection, and its exceedingly chronic sphacelus, are so many adverse conditions to an acute disease. Perhaps one answer to this is, How does inflammation at one time assume a chronic form and at another an acute form, one lasting years and the other only a few days? But it must be borne in mind that, directly as is the intensity of an epidemic, so is its acuteness, and as a rule so is its infectious nature intensified; and when supplemented by an acute rubeoloid disease, first showing itself not far distant from the original seats of both diseases—Ethiopia, which on one side has the Nile running through its territory, and, on the other, is nigh to the confines of Arabia. It is not difficult, under such circumstances, to conceive that the endemic peculiarities of each, when quickened and intensified by an epidemic era, favourable for the regeneration of both endemics, should, by its adaptency for each, make the nutritive focus of each assimi-

---

* In speaking of hybridity in such germs as are here called zooitic fungi, or active animal germ sarcode, the condition of sexes is not necessarily implied, but that germs have new properties and increased powers of propagation by one kind of germ feeding upon an allied animal germ, whereby activity, both as an infectious agent and as a poison, may be increased and stimulated so as to modify and change the kind of disease produced; just as larvæ in a hive are affected by the kind of food supplied, so that an ordinary larva is changed into a queen bee when perfect, and so becomes a reproductive bee. Culture also greatly affects the functions and nutritive properties of vegetables.

late to one general form of development and accelerated mode of progress, especially if both be in their primitive natures animal poisons, and out of the two produce a new disease, stamped with distinctions different from either, but marked with the leading lineaments belonging to both. This is what we find in the higher orders of animal life, when hybridity diverts nature from its pure line of reproduction. Yet, if it extends only to different breeds of the same species, the cross improves and ripens the breed, and the reproductiveness.

Let us now consider chronology in relation to Ancient Epidemics.

If there is a part of history more intricate than another, it is surely that period of the world between the fall of Nineveh to the coming of Christ. But for this part of history Clinton has done wonders, and he has done it honestly, for in his voluminous chronology he has carefully cited his authority in the original, and then gives his comment. Layard has even done more; he has rectified our chronology by raising a dead literature, engraven on stone, into a new life; and Rawlinson, Pote, Bunsen, and Renan have each in his way helped to unfold history in relation to chronology. Leaving the old chronology, where Sardanapalus of 818 B.C. is reduced to 650 B.C., we get to the more definite period of the Jewish Captivity in Babylon, at 587 B.C. Comparing this with the Olympiads, and the foundation of Rome, a tolerable index of time may be formed; as from the Babylonish captivity and backwards a tolerably connected chronology can be fixed, taking not the Book of Judges, which, in respect of succession, gives no data at all; but, taking the Exodus of Israel as a fixed period, and adding to that period 480 years, as given in 1 Kings, chap. vi., 1st verse, which was the fourth year of Solomon's reign, we get a very tolerable outline of chronology from the time of Christ to the Exodus and the birth of Abraham. Rejecting, therefore, the data supplied by Manetho as being in many points markedly in error, and in others savouring so much of the mythical, the old style of chronology by Usher is the one chiefly followed. This chronology is singularly confirmed by the genealogy of the Son of Man, as given in Matthew, chap. i., verses 1 to 17.

Let the passage be quoted, and a few remarks made upon it:—" So all the generations from Abraham to David (are) fourteen generations; and from David until the carrying away into Babylon (are) fourteen generations; and from the carrying away into Babylon unto, or until, Christ (are) fourteen generations."

It is inferred that the generation only reaches *to the birth of Christ*, and does not extend beyond it. Again, that it starts from Abraham, which includes (as is here supposed) his birth and onwards. One of the great difficulties is, What is meant by a generation? Secondly, Were the generations over this long period of time the self-same?

The period from Abraham to the Egyptian exodus was marked by a gradual decline in the duration of life; and, in the earlier period, from Arphaxad to Abraham, by a still more rapid decline in the duration of life. *But the duration of life reached its lowest ebb from the Egyptian exodus and onwards, and that standard is the current standard of the present day.* So important was the fact, and so new to the coming generation, that Moses expostulates and laments over it in a most feeling and painful manner, as recorded in the 90th Psalm, the only Psalm of Moses incorporated in the Book of Psalms. Moses there states that "The days of our years are three-score years and ten; and if, by reason of strength, they be four-score years, yet is their strength labour and sorrow; for it is soon cut off, and we fly away."

Before this period the days of man were on an average much longer, as is briefly tabulated below:—

    Abraham lived 175 years.
    Isaac   ,,   180 ,,
    Jacob   ,,   147 ,,
    Joseph   ,,   110 ,,
    Levi   ,,   134 ,,
    Moses   ,,   120 ,,

Hence the period for a generation between Abraham and Moses must be counted much longer than from Moses, or the Egyptian exodus, to the time of Christ.

It will be found that, for all the young men who died in the wilderness under Moses's leadership above twenty years of age, *forty years* was given for them to die off, some by special Divine visitations, and others by the course of nature, but all were included under the period of *forty years*.

Without here entering into any lengthened dissertation upon such an interesting subject, it will be laid down plainly that "generations" does not signify the order of natural succession from father to son, but *a given epoch or era, as of forty years*, in which time certain persons of a given descent were born in such an epoch or generation, as from David to Christ, and that certain essential connecting links were maintained to give a correct order or *line of descent*, but not all the direct successional parentages, as in a State peerage.

Again, 100 years is the period given for the Abrahamic generation, or that period of time which, by the word "generation," Abraham would understand the duration of time signified in his own day; as from the time of his death, which was 1821 B.C., to the time of his posterity occupying the land of the Amorites, four generations extend their duration into time, or that in or during some time of the continuance of the fourth epoch of time or "generation" his seed should possess the land of the Amorite. Now, taking Usher as our guide, we find that Abraham was born in 1996, and *died* 1821 B.C.,* and Israel entered Palestine under Joshua 1445, which gives a period of 376 years, which is at the latter end of the fourth generation or epoch of time from the time of Abraham's death; and that in the fourth generation Abraham's seed should possess the land of the Amorites was, in Abraham's time, most intelligible language to the great progenitor of the Hebrew nation, a nation the most remarkable the world has ever seen—remarkable in prosperity, in adversity, and in unheard-of hardships; and excelling all people upon earth in their pertinacity of adherence to the traditions of their fathers, and in ignoring the study of that Word upon which they maintain their traditions are based.

If we apply the above remarks to the three fourteen generations given, we shall obtain the following results :—

From the Egyptian exodus to Christ there will be thirty-seven generations, each of forty years' duration, and five generations each of 100 years' duration. Then add these two together, and it will give—

       37 generations of 40 years each are equal to   1,480
        5 generations of 100 years each are equal to   500

Total 42 generations, or three 14 generations...Total 1,980

Kepler gives the time of Messiah's appearing as six years, and not four years, before the common period of reckoning, which is the time here adopted, for reasons too lengthened to be given in detail. Therefore, to 1980 add   6

                                                  B.C. 1986

For the generations in Matthew have nothing to do with

---

* Read carefully Genesis xv. 12—21, especially the 15th and 16th verses.

our starting the era of A.D., either before or after any fixed period in our chronological systems.

According to Usher's determination of the birth of Abraham as B.C. 1996, a variation of ten years occurs, and also a variation occurs in the Egyptian exodus of B.C. 1491, which, by adding six to 1480, makes it 1486, or only five years short of Usher's assigned date, which, in so long a period, is remarkable, and a very independent corroboration of the general accuracy of the early chronology of this part of the world's history; and also remarkably confirms the time of the reign of Nebuchadnezzar and the Jewish captivity, which commenced 587. For if we add six years to 560, the length of fourteen generations of forty years each, we have the date of 566 years, or a deficit of twenty-one years; but it was eleven years later than 587 that the entire captivity was carried into effect, which would bring 587 down to 576, or only ten years longer than the period of 566. Yet this does not entirely embrace what the passage quoted omits to say, for whether carrying away into Babylon signifies the very commencement of the captivity, or some portion of that time in which they were placed in captivity, is somewhat uncertain; for Matthew, in his genealogy, referring to the time of the Babylonish captivity, expresses himself in the 11th and 12th verses of the first chapter somewhat indefinitely, and so far indefinitely that, if we were to read "from or *about* the time of the carrying away into Babylon," it would be equally as correct as reading "from the carrying away into Babylon." He remarks that "Josias begat Jechonias and his brethren *about the time they were carried away to Babylon.*" Adam Clarke makes some learned remarks about this passage, for which see his well-known "Commentary."

From the foregoing remarks, it is not unsafe to conclude that the sacred historian signified that fourteen generations gave a *close approximation* to the real date of the commencement of the captivity, *but not its exact date;* which, in so short and comprehensive a chronology as that of St. Matthew's, is most important, and strongly confirms the accuracy of the outline of time given by Usher, from the birth of Abraham to the Egyptian exodus, and again on to David (the latter end of whose reign was in the beginning of the 28th generation from Christ), and from David on to the captivity, would be the first half of that period, and the latter half from the captivity on to Christ. This, in difficult determinations of time, gives an accessory to our means of determining certain difficult dates, which heretofore has not been brought into requisition.

We now leave the subject of chronology, as a science, to apply it more directly to practice.

Petau Petavius, in his "De Doctrina Temporum," gives the date of 767 B.C. as the time when a plague spread over the whole world, as then known—some give 800 B.C., and some a later date; but all may be speaking of the self-same epidemic, traversing different countries at *nearly* identical times.

According to the view maintained in this short dissertation, to 103 B.C. add 640; there will result 743 B.C., which would be the time of the commencement of a new epidemic period. The same would be slowly dying out towards 200 B.C. This date is not far from that of Petau Petavius's, the celebrated monk, as already given.

If from this date, or 743, we date back to the Egyptian exodus, we have 748 years, or about 108 years more than the epoch of 640. From the Egyptian exodus to the Deluge is about 850 to 790 years B.C.

From the Deluge to the Egyptian exodus dates from 790 to 860 years, as expressed in round numbers, and from the exodus to Christ 1491 years, the half of which will be about

745 years, or close upon 743 years, the time which is here fixed as the *commencement* of the 640 years serial changes or metamorphosis in the epidemic order of manifestation, or modifying the type of diseases.

The period of blending, and of pestilences attaining their highest destructive powers, appears to be *after* the completion of the first 200 years, and before the last 150 years are reached, of which the Black death and Athenian plague are illustrations; but the first outbreak appears to be usually not quite so widely spread, but, within the area it reaches, quite as destructive as at any future time of its continuance.

Having made these few preliminary remarks, it is a matter of much interest to observe that from 750 years B.C., on to the coming of Christ, save in the earlier periods, where we take the Bible as our guide, we are coming in contact, for the first time, with *tolerably reliable data* from which to measure the march of nations, and the erratic forms in which science and civilization spread their not very humanizing mantle over the inhabitants of the earth.

Nineveh, Babylon, Egypt, and Persia were either at their zenith, or else close upon decadence, at the beginning of this period; and Greece both rose to, and fell from, the height of her greatness, and Rome, as a republic, attained to the height of her power during this period.

It is worthy of remark, therefore, that from 750 B.C. on to 60 B.C. *we find no record of leprosy anywhere*, but Manetho, who was a priest, or Egyptian magus, or something akin, writing in the fourth century before Christ to Greeks, in his attempt to calumniate the Jews as having brought leprosy to Egypt, and left it a nation of lepers, gives us the only proof we have that, *during this period Egypt was the great hot-bed of leprosy*, in whose territory was always to be found the elephantiasis of the Greeks.

For, in accounting for its prevalence in Egypt in his own time, he has to reproach a neighbouring and heroic nation, who were almost too few to be reckoned upon, from a political point of view, as being worthy of notice in the scale of nations; but, to remove from Egypt the stigma of leprosy, the Jews were a most suitable and safe butt of whom to make a scapegoat, and to graft upon them the stigma which belonged to themselves. For, of all historical data to show that in Egypt leprosy has been endemic since history commenced, this is the most unique and decisive we possess, since, whoever brought it, when it got to Egypt it is plain that *there it became domiciled.*

If, then, at this time it was present in Egypt, and absent elsewhere, how is it to be accounted for?

Perhaps a very meagre review of history upon this point may be deemed useful.

In the present day, taken all in all, there is no greater authority than Francis Adams, the learned translator and commentator upon Paulus Ægineta, published under the auspices of the Sydenham Society. He says, after enumerating a long list of writers: "We owe the earliest notice which we have of this disease (elephantiasis) to the poet Lucretius, who briefly mentions it in the following lines:—

"'Est elephas morbus qui propter flumina Nili
Gignitur Ægypto in mediâ neque præterea usquam.'"

Lucretius flourished and published his great poem between 57 and 55 B.C.

As Lucretius was a man capable of abstract reasoning, and in poetry could display the most subtle power of defining and explaining causes and effects, according to the point from which he viewed them, it is most important to observe his style.

From the lines quoted we infer that Lucretius considered

that leprosy was gendered in Egypt from the river Nile which flowed through it, and *it was never found in any country except Egypt*. Though he lived in the time of Pompey's greatness, and died long before his fall, it appears that his studious habits had made him comparatively indifferent to current events, and his early love of Greek literature engrossed his entire attention; hence his entire ignorance of elephas having appeared in the army of Pompey a few years earlier than the time of his publishing his great poem, which gives his opinion as to the origin of leprosy, quite apart from the recent incidents occurring in the East, and in so doing he assigns to leprosy an entire endemic origin in the land of the Nile.

We are indebted to Pliny, and not to Lucretius, for our first acquaintance with leprosy as an Italian disease of recent importation.

The great father of medicine, Hippocrates, never mentions elephantiasis, which is the Greek name for the lepra of Egypt. For anyone who desires to examine the matter carefully will find that the complaint called leprosy by the Greeks was a superficial squamous disease, while elephantiasis was that universally malignant disease which first beginning from within worked outwardly; and, after the system was more or less subjected to its power, then began the external manifestations on the skin, in the form of sores and scabs, and white patches and nodules, etc., etc.

Again, that leprosy during this period was limited to Egypt only is remarkably confirmed by an almost unsuspected historical coincidence, which is worthy of the most careful consideration.

As Grote has shown, the Greeks borrowed little from others, and gave abundantly from the self-creative genius of their own independent mode of thinking and examining all matters about which they wrote. But whence came they to use for a scaly and whitish eruption, cf a somewhat intractable, but by no means dangerous nature, the title of leprosy, or lepra ? as leper zeber, or lepra zebra, are words, no doubt, of ancient Coptic origin; and whoever used these names, lepra zebra, etc., identifies the people as being intimate with an ancient race which, through every change of dynasty, has retained its place as the poor bondsmen, amongst whom one general complaint has adhered from century to century through a long series of ages, and among whom the original name has also retained its place; and all who have lived there long have become familiar with the endemic antiquarian lepra, from contact with the old Coptic or Egyptian race.

In the book of Maccabees—or, according to the Douay version, Machabees (I. Machabees xii. 1—23)—we find an account of the Jews under Jonathan writing both to the Romans and the Spartans, the latter of whom they claim as " their brethren." It is, therefore, evident that the Lacedæmonians had in them Jewish blood, and both recognized the other as brethren, which is here quoted in full for the sake of confirmation :—

I. MACHABEES XII. 1—23 (DOUAY VERSION).

"*Jonathan renews his league with the Romans and Lacedemonians. The forces of Demetrius flee away from him. He is deceived, and made prisoner by Tryphon.*

1 And Jonathan saw that the time served him, and he chose certain men, and sent them to Rome, to confirm and to renew the amity with them : 2 And he sent letters to the Spartans, and to other places, according to the same form. 3 And they went to Rome, and entered into the senate-house, and said: Jonathan, the high priest, and the nation of the Jews, have sent us to renew the amity and alliance, as it was before. 4 And they gave them letters to their governors in every place, to conduct them into the land of Juda with peace. 5 And this is a copy of the letters which Jonathan wrote to the Spartans : 6 Jonathan, the high priest, and the ancients of the nation, and the priests, and

the rest of the people of the Jews, to the Spartans, their brethren, greeting. 7 There were letters sent long ago to Onias, the high priest, from Arius, who reigned then among you, to signify that you are our brethren, as the copy here underwritten doth specify. 8 And Onias received the ambassador with honour : and received the letters wherein there was mention made of the alliance and amity. 9 We, though we needed none of these things, having for our comfort the holy books that are in our hands, 10 Chose rather to send to you to renew the brotherhood and friendship, lest we should become strangers to you altogether: for there is a long time passed since you sent to us. 11 We, therefore, at all times without ceasing, both in our festivals, and other days wherein it is convenient, remember you in the sacrifices that we offer, and in our observances, as it is meet and becoming to remember brethren. 12 And we rejoice at your glory. 13 But we have had many troubles and wars on every side ; and the kings that are round about us have fought against us. 14 But we would not be troublesome to you, nor to the rest of our allies and friends, in these wars. 15 For we have had help from heaven, and we have been delivered, and our enemies are humbled. 16 We have chosen, therefore, Numenius, the son of Antiochus, and Antipater, the son of Jason, and have sent them to the Romans, to renew with them the former amity and alliance. 17 And we have commanded them to go also to you, and to salute you, and to deliver you our letters, concerning the renewing of our brotherhood. 18 And now you shall do well to give us an answer hereto. 19 And this is the copy of the letter which he had sent to Onias : 20 Arius, king of the Spartans, to Onias, the high priest, greeting. 21 It is found in writing concerning the Spartans, and the Jews, that they are brethren, and that they are of the stock of Abraham. 22 And now, since this is come to our knowledge, you do well to write to us of your prosperity. 23 And we have also written back to you, That our cattle, and our possessions, are yours : and yours, ours. We, therefore, have commanded that these things should be told you."

II. MACHABEES, XV. 38—40.

" 38 So these things being done with relation to Nicanor, and from that time the city being possessed by the Hebrews, I also will here make an end of my narration. 39 Which if I have done well, and as it becometh the history, it is what I desired : but if not so perfectly, it must be pardoned me. 40 For as it is hurtful to drink always wine, or always water, but pleasant to use sometimes the one and sometimes the other : so if the speech be always nicely framed, it will not be grateful to the readers. But here it shall be ended."

There is some slight difference between the Douay and the authorized version of the Books of Maccabees. It is barely possible that the Apocrypha, being non-canonical both with the Jews and the Protestants, and canonical with the Catholics, that the latter sect have taken greater pains with their translation, and their version has therefore received a prior claim; as in matters of pure history each other's conceits may be quietly waived. The portion quoted in II. Maccabees xv. 38—40 is given to meet the views *of those who disregard inspiration altogether*, that they may see for themselves that, whatever sects have made of the writer, he himself had no conception of writing from inspiration, but simply as a pleasing historian and author.

At what time they left Judæa is not given, but it was a long time since they had had intercourse.

Probably it was in the days of the early kings of Israel, when they frequently lapsed into open idolatry and followed the customs of the surrounding heathen.

Ships from Sidon probably first removed them from Palestine to Greece, or Sparta, where they lost their national religion, and fell into the practices of those heathens amongst whom they lived, and with whom they probably intermarried, as a matter of duty and propriety—forgetting altogether, or not heeding, the institutes of Moses, but not forgetting the *common diseases they had* whilst *in Egypt*, and which subsequently followed them to Palestine.

Granting the accuracy of the historian's account of the nationality of the leading people of Sparta to be Jewish (and the document has never been repudiated), it is natural to suppose that, if a skin disease spread, and became scaly and white, they would *suspect it to be leprosy;* but that disease, not being indigenous to Greece, nor yet having spread by infection—there being no epidemic tendency to aid its development in that part—no real malignant or fatal leprosy would be found amongst them; but a disease—which in outward form put on several of the indications of the Mosaic leprosy—got the title of leprosy which was simply lepra vulgaris—a most troublesome disease of a very superficial nature, and not having any fatal tendency.

But, when true leprosy came, it was unknown amongst them; its distinctness from lepra vulgaris, was readily perceived, and a new or distinct nomenclature was adopted— a disease with which they probably first became familiar whilst Alexander the Great remained in Egypt; and though called by the Egyptians lepra, they gave it at once, by a wonderful practical gift, a totally distinct name, *that the two diseases, lepra vulgaris and true leprosy, might never be confounded with each other.* Hence the confusion of titles, for one and the same disease, between the Greek and Arabic writers. If then, leprosy, properly so called, had no existence *as a spreading disease* out of Egypt from 750 B.C. and onwards to 103, or 60 B.C., how are we to account for its dying out so completely; whilst in Palestine it was so well known to the priesthood so late as 808 B.C., that its rising upon the forehead, etc., of Uzziah, when sacrilegiously entering upon the priest's functions, as a visitation from God, they immediately thrust him out of the Temple; and until the day of his death he lived in a separate house to himself?—the promptness of action at once indicating the familiarity of the priests with leprosy at that time.

Taking as a convenient date 750 B.C., or thereabouts, it appears that leprosy ceased to be a spreading disease, and was unknown *beyond its own endemic region, but other diseases adverse to the spread and infection of leprosy prevailed through-*

out the world; but by that peculiar modifying influence in endemic eras, which appears occasionally to divert the channel of disease into some new and intensifying order of complaint of a transitionary character, when about midway between the commencement and termination, in about the year 430 B.C., from Ethiopia and Asia on to Greece, disease of a spreading and infectious character—may-be of a zooitic, or an animal nature—had probably engrafted itself with *latent* germs of leprosy in an acute and highly modified form, and known in history as the plague of Athens. If so, leprosy was very genially mated to a spreading and epidemic poison or scourge. And its powers of flight arose from its newly-acquired alliance, which quickened into active development a lethargic and very chronic poison; just as cross breeds for a season are more prolific than pure, as the wild and domesticated cattle of this country; but, left to their own natural courses, they lapse back into their respective primitive forms. And so, when the epidemium of the era 750 to 103 was completed, leprosy came out again under more genial auspices for its own natural development and extension; and in the next era it spread to every country bordering the Mediterranean, and made for itself a wider and better-established name, but in a very materially modified form, to that known in the days of Moses and Elisha.

The era from the Noachian flood to the Egyptian exodus may be considered as the era of incubation or induction of diseases generally, and might with propriety be called the Chronographic era of human decay; and from the Egyptian exodus to the founding of Rome was the era of the *settlement of diseases*, inherited from the great era of human decay.

From the *settlement of diseases* onward there have been, in every new era, outbursts of diseases, undergoing many shades of divergence, intensity, and differentiation in every imaginable form—fevers and agues taking the lead, as constant pest-houses in every community; next the eruptive diseases, as measles, small-pox, and scarlatina, &c.; and, thirdly, plagues, or universal wide-spreading diseases, assuming various forms and modes of manifestation, but always at first of extensive range, and then peeping in at this city, then spreading over that country; and in this town or village, and keeping up a constant state of unsettledness in every nation or city as to when it will be their turn next. The Athenian plague, the Levant plague, the Black death, and Cholera are those to which most attention has been directed in modern times, and the small-pox and cholera because they are present neighbours; whilst of the Levant and Athenian plagues it may be said that the Historian has clothed them with the imperishable monument of a masterly and comprehensively written description.

That one, the greatest of all, and which affected mankind the most, was the Chronographic epidemium, about which we can bring no contemporary history but that which is legendary in confirmation of it; the only distinctive testimony is that which is *written* in the Bible. Hence it is called the epidemium of chronographic decay, and it is thus recorded to us:—

	Lived Years.	Married, or eldest son born at
Arphaxad (born two years after the Flood)	438	35
Selah	433	30
Eber	464	34
Peleg	239	30
Reu	239	32
Serug	230	30
Nahor	148	29
Terah	205	70

Abraham	175		100
Isaac	180		60
Jacob	147		85
Levi	137		—
Kohath	0		—
Amram	0		—
Moses	120		—

During the first epidemium there do not appear to have been any very extensive changes in the political or civil conditions of man. Egypt at this time was the centre of art and science, and probably the most intelligent and powerful nation upon earth. Ancient Thebes and the Great Pyramid bespeak a people far advanced in science, and of very singular tastes; their power of mummifying their dead is not the least of those achievements which indicate a people far advanced in civilization.

The succeeding age presents no very marked change in the condition of mankind, innumerable petty warfares, and free intercourse of nation with nation, without any special accumulation of power and military skill in any one nation, saving the Jews; who, in the reign of David, rose to great military pre-eminence, and in Solomon's reign to great commercial and social importance. No kingdom was equal to it for wealth, social rights, and security to person and property; but this is the one great empire that rose and decayed during the epidemium from 1490 to 750 B.C. So far as history is concerned, any dates prior to the Babylonish kingdom or Nineveh appear to be next to valueless, saving those arrived at through the record of the *sacred text;* but as in our day, as in all other days, when that Word was read and known among the people, much contrast of opinion was held, and still is, partly because of the necessary sequences that must be drawn from a full acceptance of its contents, which are of a kind not acceptable to the great mass of those who either hear it or read it; and, again, because very few read that Book *straight through* as they do other books, but, as it were, in detached fragments and unconnected paragraphs, especially at that time, when the intellect is best fitted to examine its worth and real merit, as from twenty to thirty years of age. Hence, from these two causes, *chiefly*, the Bible is rejected as a book of high authority by many learned and able men, by which means our best authority is very much laid on one side.

If, then, in recognizing a withering and blighting epidemium which cast its pall upon the entire family of mankind, and in 800 years, more or less, reduced the duration of life from 400 years to 70 years of age, it is possible to discover that the past records of mankind leave us *a trace* of this great change in relation to the duration of life, it will be all-important—a change which, if true, must have destroyed entire species of animals whose powers of endurance, through many changes, is less persistent than those of man, chiefly from man being a *clothing animal*, and capable, in a great measure, of creating his own external circumstances.

Of this Chronographic Epidemium there are some few but faint traces, but no great attempt will be made at proving them, each person being savoured with a sufficient amount of scepticism to allow his faith in old records to be but little influenced by the general tenour which their real, or supposed teaching may suggest, until we launch into the *pre-historic age of mankind;* when thoughts can add wings to their wearied journey, and in a few short pages we can contemplate mankind as huntsmen, worm-diggers, snail-eaters, and general consumers of vermin and vegetables; feeble, imbecile, incapable of much physical exertion, and standing up to fight their own way in the midst of forests teeming with wild beasts and creeping things, swamps and jungles, and every vicissitude of weather; and, last of all, instead of sinking into petrified organic remains, rising, like "authors under difficulties," to be the lords of creation, or chief organic compounds, who, through selection, have attained to a maximum of development.

But let us halt for a moment, and learn from true records. Mrs. Marcet, in the "History of Astronomy," issued by the Society for the Promotion of Useful Knowledge, has indicated that India, China, the Chaldees, and Egyptians had attained to a very correct knowledge of the yearly cycle of 365 days. Without entering into the fractional details, or the difficulties which beset a very correct determination of time, it may be said that they had great ability to indicate the time of eclipses. It must be said that, with their defective methods of determination, *without the aid of very long observations in the same individual,* from great defects in the powers of expression of any abstruse calculation, the accumulation of knowledge and experience, sufficient to gain a correct knowledge of the precise working of the time-piece of Nature, is next to impossible, if not so altogether. But if one man, or many men in different regions of the earth, having a special aptitude for observations of this kind, lived over 150 years of real active life, man would during that time, in a good Eastern country, not only with the instruments he used to guide his observations, get very proficient, and, by dint of observations often repeated, correct the errors of his own instruments; but also, by a long succession of observations, he would, from long experience, be capable of rectifying and harmonizing defects of an early period, and so be enabled to elicit great general facts, as guiding-posts for future generations, in clear and intelligible language, without a hundred and two formularies, to correct the defects of differential elements which must necessarily creep in, but which, by a powerful synthetical and analytical process, over short periods of time, can now weigh in the balances, and the minus or plus be added or excluded, as certain defects in the method of observation may require.

This, then, is one reason why, in those early nations, and such nations as have suffered little from the change of dynasty or masters, we find handed down to their posterity very fair outlines of the data arrived at for calculating eclipses, and of dividing time by the year and the month in ancient astronomical records, and their posterity, from shortness of life, not attaining to their wisdom.

That this may not appear absurd one further corollary will be given, in the fact that in Egypt is the largest building the world has seen, and probably considerably the oldest. The means of acquiring correct data for calculating eclipses, and the cycle of the year and the month, are given in the proportions of the building itself, known as the Great Pyramid; where a correct proportional measurement of the earth's diameter is found, the basis of our yard and inch measure, and the key to all our astronomical measurements is here preserved, as in an observatory exposed to every storm, and every change of dynasty, for perhaps 4,000 years.

Mr. John Taylor and Professor Piazzi Smyth have changed this meaningless mass of masonry into a speaking monument, that displays the science of the ancients as a diamond in a casket of gold.

With regard to Professor Smyth's view of its being built by masons divinely inspired, and so over-riding all difficulties, there is a very important objection, which is the more unwillingly given, because it is evident that the learned professor in his work* desires to honour God by ascribing inspiration to the builders. To give the matter shortly, the grand residuum of all is that *our* inch measure is inherited from and preserved in the proportions of the Great

---

* "Our Inheritance in the Great Pyramid," by Professor C. Piazzi Smyth, F.R.S.S.L.E., London, 1864.

Pyramid. Its basis of numeration is five and twenty-five inches, which latter appears to have been the common standard of measurement. This standard, when reduced to 1-25th, gives us the inch as a unit, which is 500,000,000th part of the earth's diameter. In the beautiful fitting or jointing of one stone against another, the Pyramid, like the Jews maintain for Solomon's Temple, is air-tight, or nearly so, tissue paper being too thick to pass between the seams of the joints or fittings of stone against stone in this wonderful Pyramid.

To be as brief as possible, the Jews, who were long in Egypt, *did not* bring with them the unit or inch measure, and then the foot or two feet. They brought with them no pyramidical proportion whatever, but merely a convenient system for a set of rustics, who understood pasturage and, in a measure, agriculture, and not science and fine arts. The finger, the hand, the span, and the fore-arm, to the tips of the fingers, were the natural standard, and for most purposes were admirable ready reckoners, for they were always at hand. When more civilized, they got these ready reckoners more fixed and uniform as standards, as must necessarily be the case as they became more settled, and, in many respects, much more mechanical and artistic in their works and requirements as occupants of a conquered country. When settled, the cubit was fixed at twenty-one inches and two-thirds, or more. The old astronomer Greaves, in 1639, being the one whose inquiries in the East upon the cubit and the hand-breadth are now usually quoted, gives the cubit as 21.888, and the hand-breadth as 3.640; adding these two together we get Ezekiel's cubit, for which read Ezekiel xl. 5, and xliii. 13.

At the time Ezekiel gave his measurements of the Temple, usually called Ezekiel's Temple, he was, and had been, for many years a captive in Babylon. It is singular to observe that, in giving measurements for his new Temple, the standard of measurement is altered to nearly the pyramidical standard of 25, or rather 25½ inches; and it is not improbable, when greater care has been bestowed upon these measurements, the real standard of the cubic and handbreadth will give us exactly 25 inches.

From observing the nearer approximation to the pyramidical standard in Ezekiel's cubit to the old standard cubit, it is inferred that whilst a captive in Chaldea he had found a more fixed and precise standard used by the itinerant Masons, who carried on the most important works in those days, requiring the skill alike of the mason, the smith, and the carpenter. These would be the same order of men of whom Hiram of Tyre was chief at the time of the building of Solomon's Temple. This cubit or measurement Ezekiel was divinely taught *to use* in giving the measurements of his new Temple—a measure which was probably *the standard* when Hiram quarried, and sent to Jerusalem, ready dressed and proportioned, all the stones which, fitted together, made a perfect whole, without the sound of chisel or hammer, and without the need of trowel and plaster or cement, as was the case in the building of Solomon's Temple.

Hence, though the builders of the Great Pyramid are not believed to have been divinely inspired to make a standard of weight, nor for the central coffer a standard of measure, yet it is recognized as that standard which was divinely *approved of* in the Temple measurements given by Ezekiel.

But this standard, it is inferred, was known and in use in Babylon as much as it had been in Egypt in remote ages.

And how are we to explain it? Simply that the long, careful, and assiduous attention given to the movements of the heavenly bodies, but, above all, to the action of the sun's rays during the axoidal or diurnal motion of the earth, led to certain and precise notions as to the length of the earth's axis, and to the circumference of the earth's orb at any particular spot from whence it might be taken.

A rod, or two rods of unequal length, but of very exact proportions, and placed at suitable distances from each other, upon a flat, wide, and smooth pavement, and carefully marked year by year and day by day, would in time give all the essential elements whereby the length of each particular day could be ascertained; but it would be a work of enormous labour, and would require many years to rectify and re-rectify slight errors, and a very retentive[1] memory, in addition to a very large amount of notes of daily measurements, and these tried and re-tried year by year till all was reduced to the greatest point of correctness.

Moreover, to train another as a magus in all the details of his art, and to obtain a perfect mastery of its practice, would require many years of experience before he became an *expert;* but when an expert, his predictions and calculations would be viewed as from the gods.

For in those days goodness and kindness were not deemed as godlike, but concealed wisdom, and inexplicable announcements and heroic deeds of daring, which were only seen in the dark, were those which raised a man to the estimation of being in contact with divinity. Such men, being at the beck and wink of the king, enabled him to appear before his people as a half deity and some one superhuman, who treated the poorer members of his nation as beasts of burden and as slaves. It is only under such considerations, and having as a king such wealthy men as the kings of ancient Thebes or No, that it is conceivable that the most accomplished masons the world ever saw, possessed of geometrical science, and of instruments of the most perfect constructions, would ever attempt so stupendous a work as the Great Pyramid; and nothing less than the King's Royal Command would secure its erection, accompanied with instructions to found a building which should not suffer the skill and science of the magi to die out before successors could suitably take their place. "*Seeing that life was becoming so much shorter, and the time allotted for acquiring precise knowledge, from long experience, was getting too curtailed to allow of experts becoming thoroughly finished astrologers and wise diviners*, therefore the masons received the order to build us a temple of Time, and a place for a standard of measure and weight, that our successors be not robbed of their revenue, which is paid by both these standards; which done, our best deed for our posterity will be completed." Such is supposed to be the spirit of the instructions to the Masons of "The Great Pyramid," which is left as a monument of a long-lived age; but, through the onward curtailing of days, it never served the purpose for which it was designed, as a perfect and beautiful sun-dial of great use to the State, and of daily study for the magi; it has only remained as an idle monument of human decay, to speak in these latter days of the times and doings of a bygone epidemic period, which has given us but this one amazing monument of the learning and comprehensiveness of those who lived in an age of great longevity,* which,

---

* The Great Pyramid was built as an *imperishable monument* to anticipate and preserve the labours of experts in science from being lost by the rapid decline of human longevity is manifest from its site, where rain rarely falls; its entire want of architectural beauty; its singular construction for the equal distribution of weight; its perfect uselessness as a dwelling-place; the intricacy of the way to its central chamber affording secresy; its marvellous adaptation to preserve a standard of weight by securing water undisturbed from external sources, and of a fixed temperature; its dry measure being of imperishable stone, and the chamber itself of those dimensions which give as a standard of measure the earth's diameter reducible to the inch. *Secresy* is one important matter in the whole, and is certainly the best kept by transmission to a limited number, from generation to generation, as among the Hindoos; but where decay of life is so rapid, such transmission in the Chronographic Epidemic was impossible, as father and son and Preceptor would all die old men about one and the same time. Hence the need of an imperishable Temple to Science, but was in its day probably called a "Temple to Time."

though recorded in stone, is only plainly *written* in the Bible.

To many not accustomed to enter into the views and sympathies with which scientific men embrace certain leading points of great interest, the idea of building a large structure to preserve intact their own long-laboured-for conclusions will appear most absurd, and a perfectly useless outlay of money; and many would almost count Rulers half-witted who would lend themselves to such absurd notions when the whole might be done upon parchment with indelible ink, and, in a scroll, be preserved air-tight for ages, in a much more manageable and useful form. But in so measuring the weight and importance of leading scientific research, let us read the words of one of our most accomplished and able scientific writers in 1831, who lived years after to write in *Good Words* the importance he attached to the inch measure as a unit, as recently suggested by the writings of Mr. John Taylor:—

"But it is not enough to possess a standard of this abstract kind; a real material measure must be constructed, and exact copies taken. This, however, is not very difficult; the great difficulty is to preserve it unaltered from age to age; for unless we transmit to posterity the units of our measurements, *such as we have ourselves used them*, we, in fact, only half bequeath to them our observations. This is the point too much lost sight of, and it were much to be wished that some direct provision for so important an object were made.

"Accurate and *perfectly* authentic copies of the yard and pound, executed in platina, and hermetically sealed in glass, should be deposited deep in the interior of the massive stonework of some great public building, whence they could only be rescued with a degree of difficulty sufficient to preclude their being disturbed, unless upon some very high and urgent occasion. The fact should be publicly recorded, and its memory preserved by an inscription; indeed, how much valuable and useful information of the actual existing state of arts and knowledge at any period might be transmitted to posterity in a distinct, tangible, and imperishable form, if, instead of the absurd and useless deposition of a few coins and medals under the foundations of buildings, specimens of ingenious implements, or condensed statements of scientific truths, or processes in arts and manufactures, were substituted. Will books infallibly preserve to a remote posterity all that we may desire should be hereafter known of ourselves and our discoveries, or all that posterity would wish to know? and may not a useless ceremony be thus transformed into an act of enrolment in a perpetual archive of what we most prize, and acknowledge to be most valuable?"*

Save in the point of publicity, what a commentary is the Great Pyramid upon the words and clear-sightedness of one of the greatest of modern philosophers! who lived to see the verification, in recent discovery, of the importance of material measures transmitted to posterity in an imperishable form.

From the foregoing examination of epidemics, it will be perceived that a general idea of epochs is suggested; and also, since about the year 100 or 103 B.C., the general epidemic epoch has been about 640 years, and from 200 to 400 years from the beginning of such epoch there is a tendency for poisons, distinct in their specific actions upon animal, and more especially human life, to lose some of their sharp defining pathological *effects*, and to blend and to cross, as hybrids, with each other, and so produce a very

---

* Lardner's "Cabinet Cyclopædia," 1831. "The Study of Natural Philosophy," by J. F. W. Herschel, Esq., M.A.; from page 128 and note. Longman & Co., Publishers.

modified change in their *modus operandi*, as distinct and specific in their action as poisons, either in more intense and generally destructive powers, as in the case of the Black death; or of a more chronic and milder, yet equally prevading form, as in syphilis, or blended Levant plague with leprosy. If these views are correct, it is not going too far to say that since 1817, when we entered upon a new epidemic era, the blending of the two diseases will slowly but inevitably assume some new or modified form of manifestation, or run slowly into one or the other, as hybrids under new circumstances will do, but both, under a new era, in a considerably modified form, to that belonging to their own indigenous epidemic eras.*

From 103 B.C. to 2347 B.C. is a period of 2,244 years, or thereabouts, and in this period is embraced the *Chronographic* epidemic period of about 800 years duration or more; *after which period the induction*, or great variety, of diseases to which mankind is subjected took their present essentially destructive outlines and demarcations; which, as ages have gone on, have had a tendency to gradually differentiate, and sufficiently so to demand distinct descriptions and forms of recognition, properly arranged under some general class or order of diseases, as are supplied in Nosological systems.

Then, from 1491 to 746, or 750 in round numbers, is a period of 741 years. This is considered the period of the settlement of diseases, or that period in which the constant induction of new forms of disease became less frequent, and existing diseases repeated themselves, upon the whole, with greater uniformity and constancy than in the preceding epoch.

And finally, from 750 to 103 B.C., which embraces a period of 647 years, we begin to enter upon the era of more

---

* It may be well to mention that this portion was written some years ago.

ordinary and regularly repeating epidemic eras of about 640 years duration; from which time *epidemics of a specific character* took wings to themselves, and appear to have ranged in some one particular form, or type of disease, over large and extended areas of country; and, during their prevalence in any one locality, to have kept in check and absorbed into themselves the chief mortality to which, at that specific time, human beings and cattle, etc., were subject.

CONCERNING THE POISON OF EPIDEMICS.

With all our improved pathology, during the last sixty years, including Baillie, Rotetansky, Williams, Virchow, H. Jones, J. W. Ogle, Lionel Beale, S. Wilks, and a host of contemporary writers, there still appears to be a sad want of a wider view of the active moving agency in the form of infection, when such exists, in promoting the spread of epidemics. For the spread of cholera, Sir H. Holland gave us his insect theory, and now most stand by a fungiferous theory, for its active poisonous effects on the human system. This appears to be a great advance in our etiology of epidemics; and probably the theory of fungi *imbibing decaying animal matter*, and growing and multiplying rapidly, whilst such material can be obtained in a semi-humid, or liquid condition, especially from contaminated water and moist air arising from sewers, etc., gives a very accurate idea of the leading element of localizing disease, *when the epidemic condition is present;* but only when that condition is present, which no doubt gives an aptitude in certain kinds of fungi to assimilate, after a particular manner, decaying animal matter in such way as renders them poisonous to the human frame; and, when fungi have entered into the human blood, of their acting as a catalyctic, and disposing the blood to undergo rapid and great changes in a short time. If small in quantity, then the change is much slighter; and if fresh and active, and in large.quantity, then it is necessarily fatal—first acting upon the heart, and then upon the lungs; suspending in part the functions of both those organs, whilst rapid enteric, and probably also gastric desquamation, allows the rapid filtering through of the more serous part of the blood. But, if this filtration can be checked *within given limits*, and the vitality of the blood is not entirely over-balanced, the ability of the blood to readjust the disorganization by a converse catalytic power, which resides in so much blood as is still unchanged by the morbid action of the poisonous fungi within it—then restoration to integrity in the blood is rendered easier and more certain, and resumption of the normal functions of all parts of the body, including the brain, kidneys, liver, heart, lungs, and muscles, etc., is rendered more uniform, and of much shorter duration, and therefore much safer to the patient, than if left too viscid by over-draining.

But, it is probable that agues of various kinds have a fungiferous origin, as well as cholera; and, almost to speak metaphorically, cholera is the prince of agues, and is by far the most acute and the most fatal, as well as in our own days the most wide-spread ague the world has ever seen; though cholera is of endemic origin in India, and of very great antiquity, as it is referred to in the Vedant, yet perhaps not as old as Brahminism itself. For the Divine Books, or the Vedas, were probably not completed till the early part of the Christian era, and of course the Vedant somewhat later. If such really be the case, the antiquity of cholera, so far as written evidence is concerned, will scarcely be so old as some would claim it to be, and its extension beyond the Indian continent not quite so long delayed as some writers suppose, though it has often appeared south of India in the isles of the Indian Ocean before 1817, but not north of the Himalayan range.

But, in addition to fungi, is it not possible to suppose an *animal sarcode* as perfect and yet as simple in structure as the fungi (bearing in mind in all animal growth the tripartite element), and capable of transportation, as a light and almost impalpable dust, from person to person, as much as we imagine that fungi are, as evidenced in certain skin diseases which are propagated by close contact or actual touch, and much easier of conveyance than the ponderous acari, on the animal side, propagating their brood of ova, to be kindly housed in the next neighbour's hand, who by gentle contact is sufficiently felicitous to be honoured with their presence?

Such an assumption would, in many respects, assist in explaining the singular and *regular successional changes* through which many of the zymotic diseases pass, as scarlatina, variola, rubeola, and typhus, etc. And may not typhus fever be counted but a mere variety of typhoid, and many of the miliary and spotted fevers of old authors, by accepting, in a wide and comprehensive sense, the notion of an independent existence in the *materia morbi* of infection, beyond the limit of its first and endemic origin, by admitting in epidemic periods something *akin* to the phenomena of Parthogenesis, so well established by Steenstrup and Owen, etc.; whereby it would be shown that certain epidemic conditions gave to existing zooitic fungi or animal fungi—if the term can be allowed—new and independent powers of increase and extension, which, with the subsidence of the epidemic, lapsed back to their former endemic slums and rookeries? Then we should get over many outstanding difficulties in the way of the rise and spread of infection.

Such an assumption has often appeared as almost an

Epidemics. 233

essential in accounting for some of our zymotic diseases. But for cancers, consumption, etc., it is not essential as a primary starting-point that we admit an ascending scale of cell development; but the degradation of a higher cell development to some lower and yet more independent condition of cell development, which, from undergoing the transformation into a lower form, or degradation of development, especially in cancer, is more persistent in its increase of growth.

The very nature of this rapid outline, or sketchy allusion to many important matters, necessarily excludes the idea of a lengthened examination, or attempted defence, of subjects so replete with interest, and yet so obscure in tracing the data and groundwork of the assumption so concisely given; but the suggestion must be allowed to suffice for the present.

Concerning the present epidemic epoch, from 1817 to 2457, or thereabouts, it will be said that, calculating from the past, much might be suggested for the future; but there is an old saying, "Least said the soonest mended," therefore as little will be said as possible, consistent with a very patient review of the leading facts, or at least a very fair share of them, the chief source of which knowledge has come from reading the present literature of the day, of which none is more admirable for so complete an index to disease as that supplied to us through our weekly, monthly, and quarterly medical periodicals. Here and there some author stands boldly out to claim a passing consideration who appears, in a measure, more prominent and bewitching in his style, or more terse and comprehensive in his matter, than perhaps the current periodical literature of the day can lay claim to; yet even here much has first gained its way to the public by appearing in periodical literature. Perhaps such authors as

Trousseau, Paget, Greaves, Marshall Hall, Prout, Abercrombie, and Liebig may be instanced as a few out of many from which all branches of medicine have received an impulse and a kind of tincture and bias, either from the facts they have brought out in a prominent manner, or from the methods they have pursued in arriving at the conclusions they have enunciated to the world.

Added to these is that undefinable something which personal observation acquires, or sifts and analyses, which no amount of reading and study can ever supply or engrave upon the mind with half the strength, or with equal accuracy and due appreciation of the real and doubtful from the decidedly fictitious. Therefore, taking all matters into consideration, it is contended that gradually from 1823 on to 1833 blood-letting was gently on the wane, from which time, by Marshall Hall's work upon blood-letting, it got a first decisive check; from thence till 1841 it got a more decided check, chiefly from fevers sustaining blood-letting less and less. Cases which appeared to do well at the first from the bleeding, as they advanced towards their end (third week), manifested such exhausted powers of life, that fatality was decidedly greater in typhoid and typhus fever in those who were bled than those not bled. From this time onwards bleeding was gradually falling into disgrace, and after 1854 it may be said that it was, as a universal and beneficial remedy, thoroughly condemned; from thence and onwards in Great Britain it is only occasionally resorted to, and under circumstances of very mature consideration, with the greatest regard to the quantity to be withdrawn at the time.

On the Continent, especially the south of France, Italy, and Spain, blood-letting is still much practised; but the leaning to frequent neuralgia, and general proneness to disease and local inflammation in those who have been heavily bled, is now awakening a growing anxiety for the future well-being of those who have suffered from heavy depletions.

'What does this teach? Not a new theory about blood-letting altogether, but something of a wider and far more general character. The general teaching of facts on a very large scale tends to the inference that *the heart* is, in some way or other, that organ of the body which, taken all in all, is the one which is more below par in power and function in *the present epidemic period* than that of any other organ in particular.

Many have long suggested that heart diseases are becoming much more frequent than in former times; but as to actual organic diseases it is very doubtful if such is really the case.

Perhaps dilatation is getting more frequent, especially of the right side; but for London especially it may be said that the hurried rushing to railway stations, and the sudden cessation of all muscular motion the moment after arrival, has no inconsiderable amount of sin at its door, which will account for one form of heart affection. So likewise the increased leaning in London and many of our large towns to build very high has a similar tendency, and is most felt among certain classes of servants and lodgers.

But our railways, our boat and pedestrian racing, our stupid Alpine displays of courage and folly, have really nothing to do with the stamp of an epidemic bias. Such incidents would have an injurious effect in the long run in any epidemic epoch, no matter what particular bias that might take. Such incidents might be called accidental endemic acts, inseparable from the circumstances under which any community may chance to be living, and partake of a mechanical bias rather than a vital or morbific state.

Again, heart diseases appear to be more frequent by,

perhaps, ten to one than they were formerly, because formerly they were known by their remote effects, especially upon the pulse and by dropsy; but now, long before a patient can have the most remote idea of any ailment at the heart, hardening and thickening of the valves, and old adhesions of the pericardium, can be detected by the mere motion and impulse of the heart; regurgitation and unrhythmical action, when only slight, can be frequently detected long before the pulse gives a faithful index, unless it be by the symphograph; much more pericarditis, effusion, endocarditis, and many other affections of a grosser or more apparent character can be made amenable to the several indications which careful auscultation, and, with regard to site and size, percussion supply. Hence, from the mere fact of increased diagnostic powers, occasional diseases of a very fatal character in a specific organ are remoulded into a great number of slight, and, in many instances, very manageable diseases, and into a few that are very serious, and often necessarily fatal diseases of the heart.

But what is here maintained is, that we have no direct proof of the greater tendency to disease of *an organic character* in the heart; but that, taken upon the whole, the heart, as a central organ of the body, has *less power to propel the blood* throughout the body than formerly; that the hard wiry pulse, so often found in brain affections and serous inflammations, is now scarcely ever known or felt; likewise the hard, wiry, and almost incompressible pulse is to the present, or rather rising generation, a perfect myth. The frequency of serous and congestive apoplexy is now common, and not the exceptional apoplexy; whilst apoplexy from extravasation is not so frequent as formerly. The firm, organizable, and tough or friable lymph in serous inflammations is now rarely seen; it is altogether more plastic, soft, and yielding than that of former times. Also our large ovarian operations, and large incisions for different objects through the abdominal walls, are less prone to run into intense serous inflammation. The same was long recognized among the nations of India, and the negro populations in the West, who bore abdominal incisions and recovered, when we rarely ever knew of an instance of the kind. Hey, in his " Surgical Observations," gives *one* instance of recovery in a woman after violent laceration. In fact, our entire constitutional bias is more Asiatic and African than it was; we bear large and capital operations much better, and if we are not well fed and sustained, we sink and die off under them much more quickly than formerly.

Plastic operations, conservative surgery (anticipated, in a measure, by Gooch and Hey), mechanical means for arresting hæmorrhage, as by accupressure, ligature by cat-gut, and silver wire; compression in aneurism, and a host of other improvements, justly claim our admiration as so many advances in modern surgery—a science which in its palliative and curative effects has far outstripped medicine; and in nothing has the gratitude of mankind to look back upon with more unfeigned thanks than to its last two master strides, anæsthesia and antiseptic appliances. Yet, with all these improvements, the tolerance of the human constitution to bear great operations, especially those which interfere with serous membranes, has run remarkably parallel with the decrease of tolerance to the bearing of blood-letting; and the proneness to less violent and throbbing sthenic inflammation, well described and illustrated by such authors as Alison and Cooper; whilst it is to be borne in mind that many of our leading improvements in surgery anticipated ether and chloroform, or carbolic acid and permanganate of potash, and antiseptics generally.

Such men as Abernethy, Sir A. Cooper, Dupuytren, Larey, Lisfranc, Hennen, Mott, and Warren are men of a past generation, who were as daring and skilful in surgery, with many brilliant cotemporaries, as the world ever saw; but these men would have shrunk to do what we now do with perfect success, not for want of skill or judgment, but from that wholesome dread which experience gave them of the extreme proneness which serous membranes had in their time, from very slight injuries, to run into violent and fatal inflammation.

Perhaps some will say, Did they know the right use of brandy and stimulants? Query: Do we know it ourselves? Do not cases do much better with malt liquors and wines, than with the newly acquired property which alcohol gets by being isolated by the heat applied to the Still; whereby is gendered a craving for the like to be repeated, immediately upon recovery from a debauch and sickness from its presence the night before? This appears to be contrary to nature, and certainly our ordinary malt liquors do not usually create such an *unnatural craving*, and are much safer and more sustaining—and, above all, our numerous kinds of wines—than distilled liquors, however little or much diluted.

The great increase in the use of ferruginous medicines, either artificially prepared, or supplied at some one of Nature's numerous springs, tends greatly to show how much attention is given to languid action of the heart; an action, also, almost invariably present where there is much neuralgia, however much the venous system locally, in the neighbourhood of neuralgia, may be heated and congested by plus of blood to the affected part.

If, upon the whole, the force or power of the heart is lower now than fifty years ago, what ought the natural difference in the type of disease to be? This is a hard and

Epidemics. 239

comprehensive question, but surely it deserves a passing consideration. Suppose it be assumed that 22 parts of blood represent the given quantity of blood flowing through the system, and to the arteries be allotted 6 parts, to the veins 7 parts, and to the capillaries 9 parts, which together will make 22 parts? (The quantities here given are purely hypothetical.)

The capillary system is here represented as containing more than either the venous or arterial systems; that it of course delivers its blood to the veins at a slower rate of velocity than that belonging to the arteries; but the veins, having a larger capacity for blood than the arteries, the discrepancy of *velocity* in the veins is compensated by their greater cubic contents, giving to the right side of the heart an amount equal to that injected into the arterial system by the left side of the heart. But the most difficult system to understand is the capillary system, a system which appears to have some motor power plus that of the heart, and, in a measure, independent of it; nevertheless, as a whole, it acts synchronously with the blood propulsion delivered to it by the action of the heart. Moreover, the momentum with which it receives blood from the heart invariably *regulates the momentum of the circulation* running through its own finely spread net-work of tubing; but, inasmuch as the blood travels through a larger cubic space in the capillaries than in the arteries, as well as its being checked by increased friction, the momentum of force with which it enters the veins must be considerably diminished. But put the matter in another light, and then a clearer aspect will be obtained. Let the time or velocity be the same, and the size of the bodies the same, in which two hearts are placed. One heart shall use force in one hour equal to move 100lbs., and the other force in the *same time* equal to move 80lbs., the weight being divided into fractional parts for given seconds of time; yet, howsoever it is divided, if the result be the difference of 1-5th in one hour, what effect will this have upon a system whose cubic contents is greater than that of either the arteries or the veins, and upon a system which, acting synchronously with its supply of blood from the heart, yet in itself has a certain *vis vitæ* of its own of a motor nature? The lesser force will leave the blood to be more dependent for the due and systematic delivery of each fresh amount, at each impulse of the heart, upon its own *vis vitæ;* and if through cold, excessive heat, or any depressing agency in any part, that *vis vitæ* is lowered, then the amount of blood will pass tardily into the veins, and its tardy delivery will be chiefly dependent upon the action of the heart, or the *impetus* with which it arrives at the capillaries from the heart—the natural result of which will be a partial and unequally distributed clearance of the capillaries, and increased dilatation and retardation of blood in them, with much local congestion, but of a passive character, with serous rather than fibrous exudations, and, in many instances, a chronic condition of congestion, from blood in its own vessels remaining fluid for long, and there being every now and then an effort in the capillary *vis vitæ* to expel its present contents; but, being in a chronic state of dilatation, the emptied vessels will be quickly refilled, and kept distended; whilst, if the power of the heart is more complete, and congestion does occur, it is not from want of power in the capillaries to complete the task laid upon them, for the heart itself, more or less, makes sure of this, if the tubing is left uncontracted. But it is quite possible, under greater force from the heart, if anything occurs to upset or lower the integrity of any particular part, that then the capillaries, by over friction, become irritated, and contract *too much, or their contraction lasts too long*, so that arterial blood flows too freely to the contracted tubes, and a species of acute congestion and morbid nutrition sets in, which altogether interferes with the local integrity of the part, and gives us all the results known as acute inflammation, with certain alteration of structure in the part affected.

Now, it is to the former condition of passive congestion, without much alteration of structure, saving in capillary dilatation, that attention is chiefly directed.

It is, then, this state, that in our day exists so long without running into all the sequelæ of true inflammation, that is so often witnessed, and in no organ so frequently as that of the lungs, in which, through sounds, the various changes can be traced from month to month and day to day if required.

Let brief mention be here made of chronic congestion of the lungs, so common since 1849, and which is so frequently confounded with consolidation of the lungs from pneumonia, and with extensive tubercular deposit; and which, under change of air, to the delight of the physician, and sometimes to his great astonishment, and to the advantage of the patient, frequently ends in most perfect recovery, especially if, *for two or three years in succession*, a change of air for two months is insisted upon, for fear, through sameness of atmosphere and duties, the system should become enfeebled, and as a result a relapse should occur.

Upon the whole, this affection does not appear to have obtained that share of careful attention which its importance appears to deserve.* It is very common, but mostly ends in phthisiss pulmonalis, or a very lingering type of pleuro-pneumonia. The indications of the latter rarely set in before six to twelve months have passed over, and the fatal

---

* Since this was first written great attention has been paid to chronic pneumonia, and different forms of consolidation of the lungs, beside that of tuberculosis pure and simple; but it will do no harm to record matters as they have presented themselves in the order of personal observation.

end may be deferred for three or six months longer; some cases are even still more prolonged.

A patient comes, say, a young man of twenty-five years of age. He is not himself—fancies he is thinner; keeps at his daily occupation as usual, but gets very tired by night, is fresh again after a night's rest, and goes through morning duty pretty well, but flags much towards evening. The appetite is fair, but the relish for food is less than usual; he smokes, but thinks it scarcely suits him. Here his troubles end, save, perhaps, he has some indigestion, and, may be, some palpitation, with slight dilatation of the right ventricle; and if very great nicety be used, though the right auricle is naturally possessed of the greatest share of irritability pertaining to the several cavities or sections of the heart, yet it will be found, after careful examination, that it is irritable in excess over its natural standard.

The *upper part* of both lungs antero-posteriorly are carefully examined, and the pseudo-complaining patient passes muster, the respiration being perfectly natural, both in regard to the character of the respiration and time; the pulse is 65 to 75, rarely below the first or above the latter, unless nervous under examination.

Quinine, iron, alkalines, and vegetable bitters, or some agent to act upon the liver, with instructions to avoid fatigue and cold, amount to the whole that is done. In London patients go and see some new face about once a month, if they fancy they are no better. So, after being seen once or twice, fresh medical advice is sought; the same category is gone through, saving that the patient is most particular to mention that *every morning* dark phlegm is expectorated. After it has been dislodged the cough ceases, and scarcely ever returns during the day.

Occasionally an idea is entertained that there is a deficit in the quantity of urine, which is really the case in rare instances; it is very rarely albuminous, and if so it is transient and small in quantity, but enough to be able to say it is decidedly appreciable to other eyes than the initiated. Being annoyed at the ill-success and progressive weakness, and slight but continued wasting, the patient returns to his old medical adviser; fortunately he has taken a recent cold, and much mucous rale and bronchial roughness is perceived in the upper part of the chest, but the expectoration is frothy and not very tenacious.

He gets better, and now he has been trying one and another for six months; his bronchitis is ended, but he evidently continues wasting, and assures his medical attendant that he brings up more phlegm than ever, and every morning he fancies he will be choked from the difficulty of its rising; and for the first time complains of the soreness of his sides, but usually the right side most; for this congestion is far more frequent in *the right than the left lung*. The patient is now examined towards the base of the lungs on the sides, behind, and anteriorly. As a rule, the *side and anterior portion of the lower half of the lung* is affected; and now there is a noisy, crackling crepitation over the affected part, but for the extent of surface no great amount of fever or heat exists generally, unless there is a very irritable heart, when there is usually more fever than with a quiet heart. The expectoration is not rusty nor streaked with blood, but it is becoming more finely frothy and tenacious. Soon suspicious signs of dulness in one or both lungs *appear in the upper part*, and it is not long before patches—one, two, or more—are found, where the respiration is not quite equal, rather more noisy, and yet no true crepitation nor mucous rale. It is the sound of air running over hard and non-elastic surfaces; tubercle is rapidly depositing, and in a few months it runs through softening and elimination, leaving one or more vomicæ to mark the presence of common consumption.

During this time the lower half of the lung appears frequently to improve, only to take upon it a new phase, and before the fatal result tubercle and vomicæ have found a place in the lower half of the first affected lung.

If tubercle does not appear, the chronic congestion appears to gradually extend upward; constant, slight, and, now and then, sharp pleuritic pains are felt; and the continuous dull surface, with distant crepitation, announces the presence of pleuritic *effusion*, and with it the constant desire of the patient to be raised in bed.

Neither does the evil end here, for mostly the left lung becomes involved in the same congestion, and, in rare cases, there is effusion into the pericardium, but rarely announced by distinct pericarditis; there is a dull, slow, and oppressed movement of the heart for perhaps a fortnight to a month before it appears, and a tendency to faint from any exertion, but the heart remains of natural size and rhythmical; at times a case will present itself without any valvular affection, and yet be very unrhythmical. The effusion shows itself by increased difficulty of breathing, a feebler pulse and beat of the heart, and a most distinctly-wider surface over which the palpations of the heart can be felt, which percussion more precisely defines. Dropsy, of course, in such cases, ushers in the last phase of this disease.

It is deserving of the most careful consideration that in the earlier stages of congestion, strange as it may appear, percussion gives an amount of resonance totally unaccountable; but in two months later no complaint upon this head can be made, as it is dull enough.

How, then, is it recognized at all? It is rather singular. Upon examining the lungs from above downwards, the intensity of the sound by auscultation gradually diminishes as the base is reached—indeed, before it is reached it is, in a healthy state, almost inaudible; but in these cases of chronic congestion the graduation stops suddenly, usually below the the fifth or sixth rib, and is reduced to a fine, thin, soft, and equal breeze, neither distinctly murmur nor distinctly bronchial, but soft, equal, and finely breezy, and from this point it rapidly gets less and less distinct. This is the beginning of chronic congestion.

When it is non-resonant under percussion, and no sound can be elicited by auscultation, how then is it to be distinguished from pure pneumonia, or consolidation from an old attack of broncho—or pleuro-pneumonia, or from syphilitic infiltration, which may in rare cases begin at the lower half, though this is not usual? This is a point of great difficulty at first sight; but there is a very simple method, which can be given in few words.

A given portion of lung is apparently perfectly solid. Auscultation gives no sound—percussion gives a dull heavy thud, and not the distinct pitch of tubercle. What is going on beneath? Request the patient to take several *forced* inspirations, and during this time keep the ear, or stethoscope, against the chest; if it is consolidation from inflammation there will be no result, or only a few unequal harsh, half-crackling, bronchial pipings; if syphilitic infiltration the same will be heard, but somewhat freer in certain *portions or patches* (besides which, there is a great deal of previous history to guide in syphilitic infiltration); but if the infiltration is *recent* there is a harsh breeziness, and dry piping under *forced* respiration. But where there is true congestion there will be, with forced inspiration, *very fine, equal, and soft piping,* which is invariably present in all cases of pure congestion of much intensity. If there is a spot where it is entirely absent, there will be found, as the case goes on, some indications to show that change of structure was the real cause of the absence of all sound, most likely lung apoplexy. By forced inspiration is meant several efforts to fill the chest with air by a strong *voluntary* effort.

It will be said, Have not several of our societies of late years given very full details of cases of chronic broncho-pneumonia, and also of acute and sub-acute pleuro-pneumonia; and does not the Pathological Society, in its "Transactions," give very careful reports of autopsies bearing upon this very point in a very elaborate and careful manner? Why, therefore, enter into its initiatory indications with so much form and detail? Everybody knows it must have a beginning, and also everybody knows that no two cases are exactly alike; therefore, such a parade of details is totally unnecessary.

Here and there one may be found who has examined this lung affection with much care and exactness; but for its early manifestations there is scarcely a full detail yet given, and for its full and careful description there is still an urgent demand from the pen of some one familiar with the disease, and able to do it that justice its importance demands. In the meantime, this rough outline is given for those who have not carefully examined it. But the important point in treatment rests upon its *early detection;* and its total painlessness, the equable breathing of the subject of it, and in its early stage the very frequent clear resonance of the chest on percussion, the frequent absence of cough and the *presence of dyspeptic symptoms*, with, occasionally, a very excitable heart—all of these tend to lead the mind astray from the real nature of a disease which, in its last stage, if phthisis does not appear, yet itself is about equally fatal as that destructive disease.

Again, it will be said, Why should one man be familiar with the commencement of a disease, and another equally up to the mark, or very much superior, should altogether fail to detect it in its early stage?

Abernethy sighted off his patient and diagnosed his disease without scarcely a word, and many have tried to imitate this real or supposed acuteness in a very superior and accomplished surgeon. But it is good enough for men of slower thought, and less prescience, to compass an end by much more common tactics than sight alone; and, to very dull men, the aid of two or three senses is called into requisition to make them ordinarily safe in careful diagnosis. From a habit early contracted (by reading in the *Lancet* one of Arnott's lectures), no patient, for more than twenty-five years, has ever been prescribed a dose of medicine without inquiries about the head and downwards throgh the entire trunk, and *invariably* examining, in some way or other, the heart and lungs, and the lungs from apex to *base*.

In this manner many a latent heart and lung disease has been detected where there has not been a single symptom which the eye could recognise, or the patient has mentioned, which could lead to the supposition of the slightest flaw in either of these important organs, but which auscultation or percussion has detected in a moment.

The only way to prove many of our morbid changes to have this or that origin is not in the autopsy, for that is the end of the beginning, but it is to observe the disease from its first deviation from health, and from that point to slowly mark the successive inroads from health to functional and then to structural changes; by which means a moderately correct history of the pathological changes in their order of succession and intensity may be obtained, much better than by the most exact descriptions of diseases given subsequently to their having arrived at a fixed stage of development, or from the most accurately conducted autopsy, aided by every modern appliance which chemistry, light, or micrology can afford.

If, then, in giving a description of chronic congestion of the lungs, and one which very rarely presents itself as an acute disease,* a difference exists to those usually given in works treating upon lung diseases, one apology alone can be given for the apparent difference, and it is this—that, as a rule, the lungs and heart are not subjected to ausculation if no symptoms are apparent of some chest affection, either from direct observation or from some remarks made by the patient; but the more uniform habit of examining *all patients* by auscultation in a rapid manner, and more carefully, if anything in the rapid form of examination seemed to indicate some error in function or sound, would often lead to more accurate knowledge of the course of a disease as it passes from beginning to end.

Whilst speaking about chronic congestion in man, which is also found, as before said, in an acute form, but much more rarely, the fact of pleuro-pneumonia in cattle observed chiefly since 1846 in this country must not be entirely overlooked. The disease is still present, but in a very modified condition, compared with that in which it presented itself at the earlier period.

Milch cows suffered mostly, and in London it would last from ten days to twenty-one, and now and then for six weeks. It was essentially the same in London and the country, but decidedly more acute in the country.

In the earlier part of this affection the trachea and larger bronchi were affected from the commencement of the disease, and cattle would cough up long masses of organized lymph or false membrane, similar to what is occasionally coughed up in cases of croup, only much longer and more massive.

---

* For the acute form, now much modified since 1841 in Ireland, see the *Edinburgh Medical Journal*, October, 1856, p. 360.

The heart, liver, and kidneys, especially in the country, suffered almost as much as the lungs and pleura.

Much attention has been given to this disease, both in the living animal and at the *post mortem*, years ago; but having written to a veterinary surgeon in the country, whose knowledge of the disease and extensive opportunities of witnessing it on a large scale give great weight to everything he says, his description is preferred to that of limited personal experience.

As the letter of inquiry included one or two further particulars in reference to epidemics, their introduction into the reply given will be easily understood.

March, 1851.

Dear Sir,—In reply to your inquiry, I beg to inform you that, in nearly all cattle that have come under my notice with what is termed the "disease," the symptoms are—

1st: Droops the head, appears dull, and feeds but seldom; separates from the herd; a slight but continuous oozing of saliva.

2nd: The appetite is further diminished, a slight cough, and difficulty of breathing; milk diminished.

3rd: The respiration becomes short, hurried, and difficult; the cough is hard, dry, and suppressed.

The action of the heart is violent and tumultuous, the respired air is hot and moist, the hide is dry and generally sticks to the ribs. Pressure over the regions of the kidneys and liver cause the animal to yield; in this stage there is a total loss of appetite.

4th: The respiration, which appears to be now chiefly carried on by the diaphragm and abdominal muscles, becomes still more rapid than in the former stage. An opaque film covers the eyes, which appear starting from their sockets. The diaphragm sometimes acts convulsively when the animal catches her breath, starts, and gasps as if from a stab. The body swells and acquires a doughy feel; the soft parts pit on pressure.

The breath, during the second and third stages disagreeable, in the fourth becomes horribly offensive. The stench from the excretions is insupportable, yet the animal does often exist for many days almost deprived of instinct.

The pulse, according to the severity of the disease, ranges from forty-three to one hundred strokes per minute. It is full, hard, wiry; in the fourth it is generally small but rapid, yet vibrates but seldom, and even then but little under the finger. The expectorated matter in the early stages is scanty and mucous; in the latter, viscid, tenacious, streaked with blood, and sometimes a sort of bloody purulent matter.

Post mortem. The first stomach always full; I have not seen one vomit.* The windpipe is extra vascular, the pleura filled with water, the lungs congested, dark coloured, swollen, and breaking readily under the fingers; the liver frequently enlarged to an extent that might be supposed scarcely possible, its entire surface of a dark greenish colour. On making an incision into it a quantity of dark fluid blood escapes, as from animals killed by the electric fluid. The heart presents all the appearance of violent inflammation.

The ventricles of the brain frequently found filled with purulent matter, and the enclosing parts bearing all the evidence of inflammation.

The Foot-rot.—I have known cows convey it to the sheep, and sheep to the cows. I do not know of man, dog, or cat being made ill by diseased animals.

---

* I mentioned two animals which did vomit in the letter first written.

The rot in sheep is not the least connected with the sore feet.*

The year 1838 was the first of my hearing of the mouth-and-foot disease in cattle, sheep, and pigs.—I remain, etc., etc.,

To R. F. B.   J. C.

N.B.—Mostly, if not always, false membrane is found in the trachea and bronchi.†

It may be said that in the country the disease is much more acute, or was many years ago; that the liver was rarely so intensely congested or enlarged in town as in the country, but the kidneys were much the same in both localities. The brain was very rarely affected in town. Upon the whole, the disease was more sub-acute in London, and acute in the country; and, like the tolerance of blood letting, the sthenic condition of constitution during illness, and tolerance to blood-letting, remained in full vigour much longer in the country than in large towns and London.

When an account is now read of pleuro-pneumonia as affecting cattle, the extent of disease and its intensity will be found to be very much abated, though still very fatal; but anyone characterizing the epidemic as simply pleuro-pneumonia, takes a very narrow view of the real nature of the affection.

The chronic congestion of the lungs in man from 1849 to 1853 was far more complicated than now; the trachea, and bronchi, the heart, with occasionally dropsy, and not infrequently suppression of urine for twenty-four hours, complicated this affection. But now the complaint, upon the

---

* It is a liver affection.

† In the *Farmer's Magazine*, 1851, an account of the peri-pneumonia in cattle in Auvergne is given by M. Yvart. This account refers chiefly to the lungs and pleuræ, and is fairly given for extreme cases; but I may add that J. C.'s account is much shorter, and agrees very much more with my own personal observations in 1851.

whole, resolves itself simply, in its first and second stages, into slow and spreading congestion of the lungs, and an irritable heart; and, in the third stage, into pure chronic pneumonia, with or without pleuritic effusion, and more rarely effusion into the pericardium. It appears to be identical with the chronic catarrhal pneumonia of Hedinger (*B.* and *F.*, January, 1870, p. 116).

On all hands, the diminished power of the heart's impulse in propelling the blood through the system powerfully contributes to a general abatement of all the more acute and inflammatory conditions.

It will be said that the leading epidemic of the present century—cholera—and the leading defect upon a special organ, are both placed under contribution as selecting the heart as the chief centre from whence evils spreading to other organs are powerfully aided and sustained. This is quite true, but why a fungiferous poison in the vegetable kingdom (the supposed material poison in cholera) should acquire a specific action upon the heart, more than any other organ, appears to arise from the very impressible nature of the lower forms of life to undergo changes in their development and properties from slight difference in the kind of nutriment upon which they feed; and if the direct nutriment by which they grow occasions such changes in their form and size, etc., is it difficult to conceive that some great and general agent, or force, which directs and gives form to disease, should not also give to fungi, when this force is passing over the earth in certain waves, a new power by slightly regulating their molecular changes, whilst feeding upon certain kinds of aliment—to wit, decaying animal matter and decaying vegetable matter—and in this manner each, according to the aliment upon which it feeds, presenting different properties when playing upon the human economy as a blood parasite—one tending most to this organ or special function, and the other to a distinct function or different organ?

But it will be said that the lungs are viewed by some, especially by Dr. G. Johnstone, as the centre of evil in choleraic poison. Upon this theory much may be said ; but in the midst of so much controversy one important point appears to be lost sight of—namely, that our physiology, in relation to sympathy between the heart and lungs, is scarcely sufficiently recognized. A man is *very weak;* he goes upstairs ten steps, each five inches deep, and he breathes rapidly and oppressedly. Another has *dilatation of the right side of the heart*, but is not wasted in flesh ; but let him do the same feat, and he breathes just as bad. A third has a large frame, and no excess of fat, but altogether a *feeble* heart ; with the same feat he has the same difficulties.

The cause is said to arise from either a deficit of blood to the head, or unequal distribution of blood to the lungs. But the change in the breathing is so instantaneous and so very decided, that the changed quantity of blood in the lungs in so short a time cannot possibly affect it, and certainly it is not from *increased* quantity in the lungs, as in congestion ; for it is from very carefully repeated observations, running over many years, that the conclusion is arrived at, *that the proportion of blood* in the lungs between rest and motion, or walking on flat and rising ground, as the true factor of the hurried respiration, is a pure myth. Why the heart suddenly increases its rapidity · at both ventricles, when the body is raised, belongs to the function of the heart alone; but why increased action in the respiratory muscles, and also increased respiratory action in the air cells is superadded, is truly a reflex excito-motory act, and of such a nature that it is impossible to affect the heart specifically in its motion without the entire respiratory system going along with it; the heart, upon the whole, acting more quickly on the lungs than the lungs do on the heart. This, in some measure, may account for the diminished respiratory function in cholera, though the heart is the first organ markedly depressed.

Leaving cholera, it may be truly said that neither does the *work of sympathy* stop at pure function, but it acts oddly as a nutrient disturbing element; for it is by no means limited to rheumatism to have metastasis from the heart to the lungs, as pneumonia, and back again to the heart. This *metastasis* every now and then occurs in mere chronic congestion of the lungs. And, without attempting to be in the most remote manner personal, we have known physicians of deserved reputation each scouting the diagnosis of his professional brother, from one saying one month that there is *intense congestion of the lungs*, the next month the other saying that there is *no disease of the lung whatever*, but there is *enlargement and hypertrophy of the heart*, and the feet beginning to slightly pit, the sequence of a progressive disease of the heart of one or two years' standing. But, on the days both examined, both conditions were present in the manner indicated, minus a proper *history* for two years' valvular disease or narrowing of the aortic orifice.

Such cases are very tiresome, and, what perhaps is worse still, the prognosis is so very annoying, for usually a fatal result is hinted at, if not directly affirmed. But, having now seen seven of these strange cases, they have served quite as a caution against hurried prognosis, for all ought, upon scientific grounds, to have died; but, according to an old Roman maxim—"divide and conquer"—all the cases in which this metastasis could be well proved recovered after a very trying and lengthened illness, and entirely contrary to expectation in the first three cases.

Having diverged from the more direct subject of examination, let us return back to the present epidemic epoch. It it be supposed that an idea is entertained that the great change in vital manifestation is shown only in the heart, as a leading centre of disease, such a notion would be very incorrect; for though the nutritive processes of the body, in its *molecular cell growth*, is by far too intricate a subject for existing scientific appliances to reduce to anything like an appreciable matter of fact, yet an entire survey of animal and vegetable stock, reference being had to our cultivated plants and our domesticated animals, will lead to a somewhat singular conclusion.

Since 1817 few of our species or kinds of potatoes are in existence which were favourites with the growers before that time, or even as late as 1832. The old kinds either are subject to the existing potato disease of 1846, or they are less productive, or are uncertain in their yield; hence new kinds have taken their place.

The same stands good of our strawberries, gooseberries, and more recently of our currants and raspberries.

The apple of Henry VIII., and of the Stuart dynasty, stood the test for generations, and perhaps apples of a finer flavour, or trees of finer growth and of greater productive powers, it is impossible to conceive; but now they are all either dying out, or from woodiness of flavour, or too great tartness, are entirely rejected—nay, their meagre productive powers are too scanty to view them otherwise than cumberers of the ground, and fresh grafts from old stock are anything but satisfactory, on the grounds of meagre productiveness.

Of course, pears, plums, and cherries are placed in the same general category. The cherry tree, though a tough and enduring wood, and very productive, yet the fruit is now slowly giving way to the introduction of new and better kinds; at least, if in flavour inferior, yet in size or productiveness superior to old stock.

Our glass or frame products scarcely admit comparison, for in the eighteenth century they were cultivated by amateurs as luxuries in this country, but now as necessary appendages to ordinary gardening, and also to supply a constant and increasing demand in the market.

Our poultry, cattle, sheep, and horses have undergone an equally remarkable transition, and probably ere long their food, in the form of cereals, roots, and bulbs, have undergone, and will undergo, an almost similar transition, down to our grasses and clovers.

Here we are opposed by men of science. What has crossing done for horned cattle? Observe the Teesdale short-horned cattle—first improved from Holland, again from our native wild stock, and thirdly from crossing with certain of our native domestic breeds. Is not crossing a direct improvement, purely the result of observation and close induction?

Again, is our racer the same now as in days of yore? Has he not more stature, muscle, length of stride, and, for age, much greater endurance than a hundred years ago? And has not Godolphin sternly held our blood pure, though the Barbary breed was blended with his at the start? Granted; and all subsequent crossing. But here has not the crossing been, not with fresh imported pure Arab, *which is much needed*, but with different removes of pedigree from the same great sire, Godolphin?

Our roadster, hunter, carriage horse, and draught horse—are they not all at this day in a state of transition? What breed, or cross of breeds, stands the chase the best? What kind stands our roads without shaky fetlocks and faltering knees? Are we as yet settled with any special stock? Have not all points to be selected, interspersed with some abominable flaws, which in the sale ought to be kept strictly in the background, for fear the purchaser will count him disqualified for future stock?

Which of our breeds are settled for the end for which they are wanted, even at the present day? Have not some disqualifications presented themselves that, in the eye of the breeder, are quite within the range of possible exclusion which as yet have not been surmounted?

If so, what does it say but that, for our own day and generation, the cross that will yield us a *permanent* species for many generations, which will not be perpetually going back to this fault of the dam and that fault of the sire, is not yet begotten?

But that cattle and horses do in time so cross, that some permanent stock is produced that meets existing wants, and retains its efficiency for many generations without returning to one or other of its pristine progenitors, is beyond all doubt a positive fact, as certain well-known breeds in past generations abundantly prove.

But it must be observed that chemistry, applied to agriculture, and to improved productive soils, and their contained nutrient growths, is altogether a problem of the greatest utility to man, and also of the greatest scientific difficulty; and kinds of crops, in relation to quality of breed, are of the greatest importance to the breeder and producer of the present day.

It will be said that, a queen bee being dead, the hive selects a successor in the larval state, which by the rest is so fed and nurtured that the size and form are marvellously altered and enlarged, and the ovi sac is enormously developed, and she ceases to be a neutral, but becomes the matron of a large and busy population, which, swarming, populates a fresh hive, and doubles the wealth of the lucky owner. This, then, is instinct's chemico-vital teaching upon food. But how by *science* are we to change the very fate of Nature, and to have on one acre a supply of food for twenty cattle instead of one, and those twenty shall be in all points superior to the solitary animal that fed upon the husks that Nature yielded of herself, as compared with the luxuriant products which labour, *directed by science*, freely supplies to the man who is diligent and wise in all his undertakings? Is it, then, here that we suddenly break off from mere change of breed to the facts and blessings of science?

But to return. What gave rise, more than a hundred years ago, to our aiming at improvement by crossing? Let it be suggested, in reply, Was it not because the breeds of a past age were fast lapsing into some form or other of the pristine stock from which they first originated, and with relapse all the benefits of some anterior cross were gradually bringing out the defects of successional purity; whilst, on the contrary, the accident of cross, undesigned by the wild cattle, had in itself given rise to the suggestion that purity only led to the onward tendency of defect? Hence, suitable crossing, attempted by enterprising men, was adopted to counter-check this downward degeneracy, out of which suggestive influence (or, according to Bain and Carpenter, unconscious cerebration or brain secretion of a good work instead of a vile and wicked work) arose a new stock of sires and dams, the short-horned breed, whose destiny is for some time to come to affect the stock of cattle in all lands.*

It will be said, What about the horse? The answer is, that the crossing of all breeds has altered, and in some instances greatly improved the horse, for the draught the

---

* Upon this matter consult "Youatt upon Cattle," 1834, by the Society for Promotion of Useful Knowledge.

least; for the road, the chase, and the cavalry the improvement has been considerable; but for the race, saving where feeding, stabling, and management are concerned, but very little has been done, though the old stock of the eighteenth century is not perfectly adapted for the present century, and crossing would probably change the qualities of the racer much, and make him, with our recent Royal Turkish stud, a hardier, stronger, and more enduring animal than he at present is, but probably not a fleeter or larger horse. These latter qualities, no doubt, are greatly the result of food, air, warmth, and well-regulated exercise, with abundance of ease during the breeding period. The amount of rest and gentle open air exercise, without too rich feeding in the time of breeding, is essential in improving the breed of cattle and horses.

It is too much to enter at length into any floral products, the decay of old kinds of plants and shrubs, and the adopting of new and varied kinds of the *same* species. The same may be said of crossing and improving our poultry and aviary generally. Nor yet is it quite the thing to examine our forest trees, as the oak, the beech, the elm, etc., etc. These, of all other things, with our higher mammalia, in their native wild state, suffer the least by epidemic changes within the memory of man; but yet it is not quite clear whether the oak is not rather on the decline in its toughness and growth, and the beech improved in both respects, during the last fifty years, in this country and Europe generally.

But questions of so great importance and so difficult of solution require observations and comparisons beyond the reach of most men to give, before anything can be decided worthy of much attention.

After these general observations, further inquiry into the subject of transition or metamorphosis, so far as regards vital phenomena, will be declined, feeling assured that in about

2,000 to 2,100 A.D. the generations then living on this earth will witness important changes in disease and its management, and probably in the spread and onward development of greater and wider nationalities, and less frivolous patriotic affectations than the world has ever yet seen.

Moreover, it is to be hoped that that vile and base demon, under the name of Patriot and semi-religion, who, in a nation's prosperity, will hurl it into fierce war—not to raise a trampled nation from its grave of social and political life, but to hurry into extravagance, debt, and frequently well-merited degradation—will be swept from the threshold of every home and every pettifogging principality and corner of the earth; and, in the place of a money-seeking or *vain* patriot, men will arise that will study and diligently apply themselves to band nations into families, and the labour of nations into the individual wealth of each member of the family community.

Before concluding these very imperfect reflections upon epidemics generally, and especially upon what may be called epochal epidemics, a few remarks will be made upon the rise and fall of nations in relation to epochal epidemics.

The scholar, the anthropologist, and ethnologist, must not look for any servile submission to the dicta of any particular school, or to the plan of detail as being fashioned upon this or that author's model of unfolding history or explaining its course and end. What is here written is for a defined object, and the generalizations given are those respecting which all can agree or differ without having their own special theories much disturbed this way or that.

Taking Acre, on the coast of Palestine, as a centre; from 2340 to 1200 B.C., civilization arose and spread to the west of Acre; Egypt, with Thebes or No and Memphis, was then the centre of civilization. When Egypt arose to a nation which absorbed commerce and science more perfectly than surrounding habitable portions of the earth is not known. In Abraham's time it was known for its resources as a granary; hence resorted to in famine.

Not far from 1500 B.C. the Ammonites, Moabites, and Edomites began to settle as a people of considerable power and of peculiar inventive genius and warlike habits; how long before is a matter of great difficulty to determine. These all lay on the south-west of Syria.

These several kingdoms, together with the Canaanitish nations, the Caphtors or Philistines, and the Syrians to the east of Acre, fell under the dominion of the kingdom of Israel about 1060 to 976 B.C.

From this period the kingdom of Israel, as a great power, rapidly declined, first through permanent division in itself, and now weakened by division, to the existing power of Egypt, and the rising powers of Nineveh and Babylon further to the east of Acre, one on the banks of the Tigris and the other of the Euphrates. These two kingdoms appear to have rivalled each other, and alternately to have been under subjection to one or the other till about 606 B.C., when Nineveh finally fell under the superior power of Babylon.

About 715, or more, B.C., Israel was absorbed into the kingdom of Nineveh or Assyria; and Judah, 587 B.C., and Tyre, 572 B.C., alike fell under the sway of Babylon. From 606 to 538 B.C. Babylon may be considered as the greatest monarchy the world had seen, whose centre was on the banks of the Euphrates.

This empire may therefore be considered as the ending of a series of dynasties, of a circumscribed distance from the centre assumed—namely, Acre, beginning with Egypt, and ending about 538 B.C. with the fall of Babylon. This, then, was the last embers of a dying-out greatness which commenced to wane about 730, and went on with rapid steps till the death-blow to Assyria and Tyre gave to Babylon the crushing centralization of all power in the East and Egypt; which, when it fell, all Asia Minor fell with it under the prowess and tact of Cyrus the Great.

Before calling the attention to Greece and Rome, let a rapid outline be given of Asia Minor and Egypt; for Western Asia and Eastern Africa, with or without Æthiopia, had a never-ending vicarious existence, until Rome stepped in to settle all differences, and, finally, to absorb them under Pompey and Cæsar.

Babylon succumbed to the Medo-Persian dynasty, founded and consolidated by Cyrus the Great about 538 B.C. This dynasty continued for about 207 years as the head kingdom in Asia Minor and Persia, when it fell, in 331 B.C., before the prowess and energy of Alexander the Great of Macedon. This latter kingdom, the kingdom of Macedon, fell under the power of Rome 168 B.C., and the remainder of Alexander's dominions fell into the hands of the Romans under Pompey the Great, and Cæsar, and their successors from 63 B.C., and ending in 70 A.D. by the fall of Jerusalem under Titus.

It may be here observed that from 750 to 100 B.C., with the exception of the little valiant kingdom of Judah under the Maccabees, that whilst Asia Minor was a centre of great and gradually divided powers, the people themselves generally became more and more servile, so far as history can well trace them, and submissive to authority, without regard to its legality or otherwise. This was the ruling character amongst both the Asiatics and Egyptians, which remains to this day.

Hence dynasties whose chief element was Asiatic fought against and hated one another, not from the daring and bravery of their soldiers so much as from their numbers, and the chance of their being led by an energetic commander, the fear of whose vigilance and severity urged them to fight far more than the contempt or even fear in which they held their foe. So the effeminancy and spirit of servile obedience, without regard to truth or honesty, were slowly but surely crushing the nationality and independency of the Egyptians and Asiatics during this epoch, from 750 to 100 B.C.

But, whilst this spirit was spreading over the population of Western Asia, a new era was coming over the nations to the north of Acre, Greece, and Rome; and, in a measure, Carthage, further to the west of Africa, partook of a popular and free or independent character.

Not far from 750, Greece a little earlier, and Rome a little later fix, by their own chronology of Olympiads and Anno Urbis Conditæ, the starting-points of their respective dynasties.

Rome lived for about 1,200 years, and ended by a pack of roughs and a small people, the Heruli, under a daring and sagacious leader, Odoacer, 475 A.D.

Greece ended by the fall of Athens, under Sylla, 86 B.C.; but her chief power fell with Macedon, 168 B.C., or 82 years before the fall of Athens, after a duration of about 600 to 680 years.

Whatever remnant there was of Rome, as a part of the old Roman Empire under Belisarius, the famous general of Justinian, the final extinguisher of Rome, as a centre of military power, was put on by the successor of Belisarius; for Narses created Ravenna as an appendage to the Eastern Empire, under the title of Exarchate, 568 A.D., at which time Rome was a province, or subjected to the authority of the Exarchate of Ravenna, from which she emerged as an ecclesiastical capital under Pepin and his son, Charlemagne.

From 750—or rather 730 A.U.C. of Varro—to 568 is 1,298 years, being a difference of about 90 years or more between the fall of Rome, under Odoacer, and its final extinction as a capital, under Narses, 567 or 568 A.D.

Greece rose to the zenith of her military greatness as a compact state, and, at the same time, was the mistress of art, science, and philosophy, from 490 to 360 B.C. During this period lived Miltiades, Themistocles, the indomitable Leonidas, Aristides, Pausanias, Pericles, Socrates, Alcibiades, Phidias, Plato, Praxiteles, and Epaminondas, with a host of other great and wise men, whose like the world has never seen since for originality of thought and perfection of composition, or consummation of symmetrical and ideal beauty.

As a great and *aggressive power*, her training under the aggressive policy of the Medo-Persian dynasty placed her suddenly on the pinnacle of fame from 336 to 323 B.C., under the command of Alexander the Great, and introduced her to an empire which, under the extent of its territory and weight of administration, divided itself into four considerable kingdoms; but that which was of the shortest duration after the division of Alexander's empire was the parent of the rest—namely, the kingdom of Macedon, as before mentioned, which ended in 168 B.C.; and the glory of Greece, the centre of learning and nationality, the city of Athens, surrendered to Rome 86 B.C. So Greece became a Roman province.

Rome, on the contrary, gradually rose from her Samnite victories to her death struggle with Carthage from 264 to 146 B.C., when, after three dreadful wars, with Hannibal, the greatest of all generals, at the gates of Rome in the second Punic war, she was obliged to submit to the degradation of a Roman province; whilst Marius, Sylla, Pompey, and Cæsar to the north, the east, and the south of Rome added kingdom to kingdom, declaring thereby to surrounding people the greatness and majesty of Rome.

Nor can it be said that the name of Rome ceased to be a terror, and the seat of appeal, to the nations of the known world till the time of Diocletian, who by associating Maximianus in the rule of the empire divided council in

286 A.D., giving to his colleague the western portion of the empire, himself retaining the east as his portion. This independent rule and division of authority led his successor to further embody the idea by material agency. Hence sprung up, under Constantine the Great, the eastern capital of the Roman world; and this capital remained, under a long succession of events, as the capital of the Greek empire till 1453, when Mahomet II. made the city of Constantinople the future capital of the Ottoman empire, which remains so to this day.

But the majesty of Rome, though marred and reduced, still retained a venerable and dignified position until the persistent childhood of Arcadius and Honorius, the sons of Theodosius the Great, by their puerile rule, neither aided each other nor defended themselves, nor appreciated those who did defend them, and so Rome lost her majesty and authority at one blow; and in less than a century after submitted to the sovereignty of Ravenna, in 568 A.D., as a dependent city, as before stated.

About this time an uneasiness began in the troubled sea of Arabian nomad life, and the land of the desert began to show symptoms of a restless tribal discontentedness; perhaps it is not far from fact to say that it began about 530 A.D. Chemistry, medicine, algebra, and astronomy found amongst its olive and dark tribes an attractive home. A blind credulity, which first led idolatry to be stamped out of their peninsula, drove them on to see in themselves a mightier power than the spreading savour of that Gospel which for three years was preached by the greatest of the apostles— Paul—in their sandy plains, and which, with occasional preachers or Jewish disciples, worked as a leaven for centuries. But, in the might of their own individuality, and in the assurance of their own Impostor, who performed all his miracles in the night and in the dark, they began, in

632 A.D., as religious fanatics and earnest disciples, to demand the acceptance of the Koran as an obligatory condition of peace from Egypt and North Africa to Spain and the South of France, all of which lie to the west of Arabia and to the west of Acre; and, to the east of the same point, from Asia Minor to Persia and India, Arabian power asserted its supremacy.

With a stoicism, mingled with a fanaticism, the Arabian nomad spread, by his conquests and his abstract sciences, a species of adventurous chivalry and philosophical indifference, with undoubted military prowess and daring, that placed the Saracen at the head of the nations of the world as a conquering power, Cordova being their Athens, and nursery of refinement and learning.

It was a species of barbaric civilization which gave to Europe its present numerals, and to mathematics all its earlier algebraic signs, which is the true source of our great advance over the ancients in the physical sciences.

The Saracenic empire, as a conquering and ruling power, lasted from 632 to 1237 in Spain, when their cousins, the Moors of Africa, overthrew them; and in Asia till 1258, when the Tartars took Bagdad. But their blood and their customs still, more or less, remain in the nations they conquered and overran.

Its duration was between 600 and 700 years, and all they have left behind them belongs to the scientific world, and the world of blind credulity in an impostor; whilst the Romans have left us as the basis of our laws their system of jurisprudence, which is engrafted upon that inimitable institution of Gothic origin, the jury; whilst the Greeks have influenced our oratory, sculpture, and architecture, and almost every department of art, science, and government, more than any other nation. Nay, the genius of the Greek nation is the admiration and the legacy of all nations, and, though dead, she yet speaks in all the sons of taste and refinement, since, by printing, her works have been familiarized to posterity.

We next come to Europe generally, the true political focus of which has stranded to the north-west from Acre, including Italy, Germany, France, and England, Denmark and South Sweden with Norway, Prussia, Poland; whilst Russia, to the north of Moscow, has taken no important part in the history of the world till the Turkish supremacy, after 1453 and later.

But who is going into the subject of the Visi-Goths of Spain, the Ostro-Goths of Italy, the Alemanni, the Saxons, Suevi, Burgundians, Belgians, Gepidæ, Scandinavians, the Norsemen, Gauls, Franks (Eastern and Western), the Vandals, Huns, Sarmatians, Finns, and Magyars? Suffice it to say such people lived, and though they are all dead now, under one name or another, in their successors, they have gone back to where they came from, as the Huns; or, through cross with other and neighbouring tribes and kindred, their names are lost but not their blood, as the ancient Iberians with the Visi-Goths and Moors; the Gauls with the Franks; the Aborigines of Central Germany—call them Kitcheners, or Earth-diggers, or by any other name—with the Alemanni and their confederates. These people, having given national customs and warlike habits, are admitted as inhabiting Central and Western Europe. They lived, and acted, and fought, and hated, and despised one another as nationalities, and as countrymen they oppressed one another in their domestic and social relations, and helped one another in their national existence.

Making one or two splendid exceptions, especially in Charlemagne, Alfred the Great, Otho the Great, and, if it is permissible, Canute the Great, and a host of smaller fry, it is scarcely necessary to say that from 568, or rather 537, there is scarcely a man in Europe of an enlarged mind, and possessing a real love of liberty, learning, or of wise administrative capacity, saving what is held as a birthright from the customs and manners of the nation or the tribe to which they belonged, that can be found from this date—537 to 1177.

Henry Beauclerk, or Henry I., is said to have had a love of learning rather than its possession, 1100, and he gave a kind of spur to learning, as historians tell us; but who were those that found him a patron of learning is more a matter of assertion than of illustration. But it is evident that his love of learning was so exceptional in those days that any man who could have composed, after the days of Boethius, anything equal to "Jack and Jill," or the tale of the "Cow with the Crumpled Horn," would have been handed down to posterity as the patron of learning, and the producer of superior rhythm in verse for the age in which he lived. Percy and Hallam, before this date (1177), can scarcely give us one verse of superior, if equal, rhyme to the two samples indicated. It was in this age of darkness that the Saracen stood foremost among the nations.

To rob the Saracen of his natural intellectual superiority, it is said that they borrowed largely from the Romans, but especially from the Greeks; and much of their apparent superiority arose from how they availed themselves of the learning of their predecessors in the republic of literature and science. Yet it must be observed that the early portion of the Saracenic conquests was marked by a people who viewed all learning contrary to the Koran as heretical, and all that was not opposed to it as unnecessary, for it contained all that man required; hence the burning of the Alexandrian Library, and the general rejection of learning of a foreign origin till a later period.

With all this against the onward development of intellect and culture, they managed to erect upon the native vigour of their own minds a system of notation and of signs, which has been the basis of our present advanced state in the abstract sciences; and in later days their knowledge of Greek writers in matters of medicine, etc., proved how ready they were to appreciate the works of other men, where the admission did not directly interfere with the arbitrary rules of their own tenets and doctrines.

But the nations which swarmed Europe, during the decay and final fall of the Roman world, were by tenet willing to adopt the faith of the conquered people, and to adapt themselves to many of their laws and customs. Yet with equal, if not far greater opportunities of acquiring a knowledge of the literature of the ancients, their works and lives remained for centuries a cipher and an untold tale to the nations of North-Western Europe, their decay in literature and learning advanced with rapid strides, and, as this epidemic period ended in 1177, thick darkness covered the face of all minds and intellects to the North-West. Italy, the focus and centre of all that was learned and great, had sunk into the lowest depths.

From 1177, and onwards for 100 years, the progress of knowledge was slow but distinct.

Within this period the pointed or Gothic architecture appeared, at the beginning of the century. Next the English minstrels, the German minnesingers, and the French troubadours began to produce a rude form of poetical composition, and a kind of ruder music. Painting, also, was more cultivated. These all indicate a slow but increasing power of intellect. But the frequent burning and persecution of heretics, as the Albigenses and Waldenses, the forbidding the reading of the Bible to the laity, and the more general introduction of auricular confession, all tend to show the rising of an independent and thinking people, while those in authority were feeling that, by this craft we live and all opposition must be suppressed, for fear that our craft should fail.

The third, or great and awfully destructive crusade, occurred at the early part of this epidemic era, or in 1188 to 1190 A.D., and many smaller ones after it; but the blind foolery which led such hosts to leave home and fight, more with the elements and hunger than with the avowed enemy, seemed to act as a leaven upon the popular mind; and many of the leaders began to reflect, and to refuse a blind obedience to the entreaties and exhortations of one man, who in his wisdom had taught them the folly of obeying his mandates, equally indicate a better tone of mind, and more correct appeciation of the merits and demerits of the teachers.

But, *before this period*, failures and badly concerted plans neither led them to see the folly of their labours, nor the utter heedlessness with which the lives of thousands upon thousands were sacrificed to the whim and caprice of mere vainglory, if such a designation is not giving to these undertakings a more honourable motive than really belonged to them. But now the tide of reflection had set in, and before 100 years were over the name of crusades was the watchword for resistance and indifference.

In our own land the first grasp at true and substantial liberty took its rise within this period, in the grant of the Magna Charta, 1215 A.D.

Having said thus much for the first hundred years, amidst all the feuds, wars, bloodshed, and chivalry, which rather disgraced than honoured humanity, no very great advance took place from this time till 1438 to 1471, when the art of printing began to be fairly established.

But the craving desire for learning had set in before the

discovery of this art. The Reformation had set in about the beginning of the fifteenth century. The great morning star of that mighty revolution of thought and independent reasoning made his appearance, under the name of Martin Luther, 1417.

The craving for tracts and works, by the great Reformer and against him, created the necessity for the more rapid multiplication of copies than writing could supply; hence came in the art divine, by human hands, of printing.

This great impulse was no doubt aided by the fall of Constantinople, and the flow of Greek literature into Europe through that incident.

From the reign of Henry VIII. till the close of the reign of George III., the spread of knowledge and the expansion and independence of the human mind, more like a god than a created being, had flown as from a fountain of troubled waters on a sea of never-ending conflict, with never-ceasing rewards to those who, in their frail barks, cast themselves upon its tempestuous waves; but, strange to say, as the journey ended, and the shore is sighted, the grand result of culminating forces has been a good landing, and accumulating security to man, with increasing wealth and prosperity to all nations, through this fountain of knowledge, whose waters are often bitter in the mouth, but in their effects sweet, and give greater security to property, and add immensely to the well-being of the human family, for the widespread increase of knowledge has tended towards power and independence.

A few names may be here mentioned from Luther, as Calvin, Erasmus, Galileo, Shakespeare, Bacon, Descartes, Milton, Bunyan, Locke, Newton, Smith, on to Cuvier, Napoleon, Pitt, Herschell, Davy, Berzelius, Franklin, Jonathan Edwards, and Kant, Byron and Goëthe.

To give anything like a comprehensive view of the vast

field which has been so rapidly passed over is foreign to the present purpose, and anything but desirable. The thoughtful reader will, in this sketchy outline, supply the deficiencies by his own ready and comprehensive memory, aided, no doubt, by many painful and some pleasant reflections. But, on the relation in which these remarks stand to epidemics, a few words may be said.

Taking as a sample the epidemic period of 1177 to 1817; we have the Levant plague running through till about 1777, when it ended in Moscow, and then it settled in its own centre between Egypt and Turkey, or Acre as a centre. In 1348 the Black death, and 1517, its successor, the Sweating sickness, gave a new form to disease, and intensified mortality by adding to the existing form of plague; whilst about 1495 to 1497 syphilis created a new order of disease, or was itself an amalgam of two previous and old diseases (as plague and leprosy), which maintains a vigorous dominion to this present day.

If, then, we take into consideration, in epidemics, from an anthropological point of view, peoples and forms of government, their rise and spread from distinct geographical areas (as the Levant plague, small-pox, and syphilis as analogues in material disease), and observe their rise, spread, sudden development of power, decline, and decay, we shall observe a remarkable coincidence and analogy between them.*

Take the Grecian kingdom. Its rise was about the beginning of a new epidemic period, or near the year 750 B.C., and Rome was not far from the same period.

---

* It must be borne in mind that in Epidemic, as well as Historic, Eras the time for development, extension, and decline, is considered to be *about* 640 years. But it is not to be understood as a sharply-defined line in Chronology, since either eras may be a few years antedated or postdated. In such cases, it may be well to view the slight variation in date as the morning and evening stars announcing the advent or departure of the respective epochs.

It rose to its greatest glory from 490 to 360 B.C. as a centre of learning and refinement, but about 400 years, or rather more, from its foundation, it launched out into a mighty empire, and then it began slowly, and in a little more than a century to rapidly decline, and in about 650 or 640 years to be practically an extinct kingdom.

This in many points bears a remarkable resemblance to the epidemic period between 1177 to 1817, especially when we consider the frequent changes of geographical centres of power in Greece, as Sparta, Athens, Thebes, Macedon, and Corinth; so the plague had so many endemic centres, not to mention its sudden accession of power in the Black death in Europe in 1348, and the Sweating sickness after.

Again, Rome more resembled the plague from 550, or thereabouts, to 1817, the earlier part of this plague ravaging all the lands on the Mediterranean, and spreading inland until checked by mountain ranges. But after 1177, or thereabouts, perhaps rather earlier, it outstepped the boundaries of its first spread, and got into Mid and Northern Europe, asserting its dominion over all other diseases during the time of its invasion of any particular city or province.

Rome spread gradually her power and conquests throughout Italy, then to Carthage, and Macedonia; and as her first period of development was reached, which would be about 90 or 80 B.C., she began to take a wider range, and more comprehensive forms of rule and discipline, in the form of Dictatorship; and from Marius on to Octavius Cæsar she raised the power of her dominion, both in resources and territory, till her sovereignty extended over Central and North-Western Europe, and from Egypt to all Asia Minor. After this period she once or twice extended her dominion towards Persia and the Caspian, but the weight of her empire enfeebled her aggressive powers.

For the first 200 years, or from 80 B.C. to 120 A.D., she reigned in glory, and acquired a high position for taste, learning, and refinement, and wonderful organization and legal administration. She now began to amalgamate in her population and customs with many of the conquered nations, and by 400 years, or more, after her new influx of greatness and power, or about 320 A.D., she had attained the chief end of her greatness; and, during the next 240 years or so of her existence as a political power, Rome declined and rapidly sank under the gathering clouds of ignorance, sloth, and bigotry.

Scarcely did the old Roman provinces find that they were free from the power and slavery of Rome, than a new epidemic period commenced, and with it a nation arose to the south of Greece and Rome, but not far distant, as a centre, from Acre — the *illiterate*, unorganized, and reverential servants to custom, which latter was the only basis of real civilization then possessed by the Arabs. But the nomad tribes of the land of the *desert*, or the Saracens, sprang as a mighty host into teachers of faith and doctrine, and admirers of abstract sciences, chemistry, algebra, astronomy, and the humane science of medicine. With discipline and courage they became arbitrators of the largest portion of the old civilized world, and that, despite all the previous knowledge, discipline, and arts of the nations which held their conquered provinces for centuries before them.

A revolution so great and so lasting, for it ran out for quite 640 years, is in itself the most singular illustration of nations rising and falling with epidemic periods which history unfolds, and in this instance very much in alliance with its own gendered epidemic diseases—small-pox and measles.

Its decay is as remarkable as its rise, and no usual

method of explaining this remarkable anthropological phenomenon has as yet been given, after the ordinary method of explaining such incidents in other nations and races.

On the threshold of its decay, and taking a wider area from the centre, Acre, fallen and benighted Europe, great in the pride of her petty strifes and interminable quarrels, begins to shake off the dust of her slumber; and, as a mighty giant, scarcely knew that she had lain down to be shorn of her strength by treacherous Delilah, till her opening eyes and the thongs of her feudal slavery warned her that she was naked, and wretched, and poor. But, with a slow and onward march, the intellect of Europe widens, deepens, and expands, till in about 1580 to 1680 the brightness of her glory and the strength of her majesty muster around her memory intellects as great as the world ever saw; and like Rome's first anthropological era, in 1817, when her first era ended, she set in a halo of greater and more commanding freedom and independent government, with peoples more enlightened and laws more equal, and, since 1600, with further extension to the East, and the West, and the South, of the principles of liberty and progress than the world had ever seen at any period of its existence.

This anthropological era, with its rich accumulations of intellectual and material wealth, it has bequeathed to the era in which we live; and, taking Acre as a centre, it is easy to perceive that in the increase of area from that centre there will be a greater, wider, and more minute and exacting form of social intercourse, which will spread and influence the entire globe more than man has yet seen, and will unite all nations and families in one great social compact.

It being laid down as a principle of civilization, from a geographical basis, that it spreads in successive increasing curves of area from its centre, which is Acre; that it began not far from about the year 1490 B.C., and has proceeded on to 1817 A.D., but we have in our day arrived at an area of circle which, in its extension, bids fair to include the whole earth; and, as a necessary sequence of such widespread civilization, in consequence of its central position by land and by sea, the increasing intercourse of mankind and the commissariats of the world being necessarily interwoven with each other, it is easy to perceive that the old centre, Acre, by the singular site it occupies in the earth, must become the great *depôt* of the increasing expansion of commerce and civilization.

But this has little to do with our subject, save as an incidental coincidence. Our chief points for consideration are the necessary effects of the more general expansion of the human mind throughout all branches of the human family.

The essential feature that most clearly manifests itself in the early part of the existing anthropological era is the strong bias in all undertakings, and in all attempts towards progress and improvements, to adopt A S S O C I A T I O N and wide-spread combinations as a basis or principle of action.

Furthermore, added to the associative principle, is the ever-increasing adaptation of the inventive genius of man to impress upon man the necessity there is for his neighbour to be his debtor for the comforts, conveniences, and safety which each may enjoy.

Among these, steam conveyance by land and sea, electric telegraph, cheap postage, artificial lighting, and above all compulsory education, which is fast entwining itself throughout every government in the civilized world, rank the foremost in aiding the rapid advance of the mutual dependence of all mankind one upon another, and each feeling he is for all he possesses more or less dependent upon the good offices and industry of his surrounding neighbours; or more distant adventurers, who ply by ship or land from distant parts for mutual exchange of goods; or, through instruction by missionaries or other friends, are brought into sympathy and fellowship as individual members of one large and variously-gifted family.

From whatever point it is viewed—open or secret, social, political, or religious—the great and grand movements in the world are now promoted, and effect is given to them, by association, and not by the wealth, or influence, or social position of any one individual. In all these movements the great centre Estate is the Press. Gas, steam, electricity, and education may all lend their aid to secure one general result, but daily and periodical literature is the nest in which all are nurtured, and in their onward movements are sustained.

But without a strong bias, which by feeding is promoted, for progress and development, the Press might work in vain, and individual members might groan and sigh under wrongs and servitude; but the individual desire for position in the social world is so great, and oftentimes repeated with a never-ceasing multiplication of adventurers in the same field of change and advance, that the stream of free and independent thinkers, from the lowest grade of intellect to that of the most multifarious and gigantic proportions, is waxing greater and greater; so that, ere this epidemic or anthropological era is ended, the supremacy of free and associative thought appears likely to encircle the entire world. Black and olive, copper and florid, or white appear from the remotest parts of land and sea, within a range of little less than sixty years, to have encompassed a circle of transition from bondage to independent, though limited, thought, which throws into the shade all efforts at general civilization which have influenced the world during any previous period of its history.

Observe our enormous armaments—our Crimean, Franco-Austrian and Franco-Prussian wars, the American Federal and Confederate war, and the Brazilian war; how, in every instance, the *numbers engaged*, as compared with like territories, for equal chances of loss and gain, in former times, are truly frightful, and the expenditure, for the luxury indulged, is wholesale; yet, strange to say, with an off-set on one side or the other of economy and benefit!

Talk of the Teutons and Cimbri in old Marius's time, who cut them up by the thousands upon thousands; and of the armies of Xerxes, against whom 300 Greeks made an abiding and depressing impression upon the half rabble and ill-disciplined troops of an Asiatic despot; what are these compared with the well-accoutred, disciplined, and carefully-provisioned troops of a minor European state of the present day? Why, they would be but food for slaughter, and as chaff before the wind for efficiency and destructiveness.

Turn to our ramifying (and—to their fellow workmen—arbitrary) trades unions; societies for sickness, death, and burial; our insurance companies, scientific and social sciences. Such, in all grades and ranks of society, is the tendency to gather strength by *association*, that even gun clubs for the destruction of sparrows have not been found wanting in certain of our young adventurers in the field of distinction and merit; and, when great fields are occupied, the tendency to association will extend itself to the honourable distinction of Fellow of the Black Beetle and Rat Club. What may we infer from this widespread and open system of association, as distinct from guilds and secret societies common in the Middle Ages? and for their very existence and maintenance each and all are debtors to the Press for publicity.

Without begging the question too far, if safety to society is the great enigma to be solved, its natural and true escape from anarchy, and its elevation to greatness and permanency, is to reject as a basis of ethics or code of social integrity *any author* who teaches supremacy to any branch or race of mankind. Equal rights and equal liberties must be accorded alike to all, so that a basis of common rights and equal law may be accorded to all ; otherwise the social advance of mankind will be blocked by unequal and highly egotistical assumptions ; and bondage of one branch of mankind under the other must follow as the practical result of such assumptions. Hence the Bible, which admits all mankind to be one family, and demands equal law, with widely different obligations according to natural gifts and special privileges, must necessarily be the basis and starting-point of true national ethics among mankind. To all these general observations, it will be said that the dark tribes of mankind are inferior to the white tribes, and the woolly-haired can never compete with the straight-haired ; nor yet the dark with the light and fair of mankind.

Such talk is grand in the extreme. But, a hundred generations back, what were the colour and features of our progenitors ? Again, a hundred generations back, what were the white people doing, and where did they live ? In Central Asia ? in the wilds of Siberia ? or on the top of the Himalaya Mountains ? It is to be feared that the Dead Letter Office would give answer—" Address not known."

You say the colour is nothing. Oh, indeed! That is nothing. Well, it is all features. Granted. The lady in Egypt had a small and arched foot, small hand, and more receding forehead. Granted. She was barren, and therefore extinct in another generation. And the Circassian ? Oh! what of him or her ? Why, when free, a nation of masters and slaves, and a father and a brother to a sister or a daughter as low in moral status and as stupid as any serfs that trod on English ground, and even baser than they; but in features and colour, why, *ne plus ultra*.

It will be said the monuments of antiquity have not been examined, and only a partial history of mankind has been taken under review. Answer: The monuments of antiquity are occasionally examined with an intense desire to get at facts—nay, now and then a skeleton and a skull; and it is surprising to find that the slave then is now the master. The Jew in chains is the chief among men for meanness, for intellect, for integrity, and for generosity. In fact, there is not at present a great nation upon earth which does not view as half-effeminate and degraded the people who once were sovereigns upon earth.

You say history is at fault; races never change. If, then, they do not change they must die out; and so history, as a record of facts, must be very defective. Well, it is true history is defective, but geography is scarcely so much at fault as the chronicler. If, then, the geographer is moderately correct, and given races leave one locality in Asia and come to another in Europe, where were the present occupants of Europe, or what were they doing, 3,000 years ago, or 2,000 years ago? Were they in the wilds of Tartary? Nay, say the Franks, eastern and western, are descendants of the remnants of Babylon and Nineveh; and the Goths, or Saxons, etc., etc., came from somewhere, but arrived on one side the Caspian before touching Sarmatia, Asia Minor, or Europe. Their pedigree is wonderful, and their deeds mighty. But how about these Teutons, Saxons, and Alemanni, etc.? Why, if they were not great *they ought to have been great*, for we are their children; and our forefathers buried their greatness in the wilds of oblivion, or in some menial and pettifogging community, so that history is silent, because the children did not honour and write for mankind the deeds they did in the past.

The upshot of the whole matter is this—that, so far as defective history gives us data, either by architectural monuments or by written statements, either the best blood of ancient races died out after a given culminating point of progress, and left the field to be occupied by inferior but more abiding blood; or, as a man at the top of the tree gets some untoward bad luck, begins to get lower and lower, till the next generation forgets his position and greatness, and the poor in the neighbourhood alone remember the habitations of fallen greatness, so a nation, once trampled down and exhausted of its resources and liberties, falls, in lapse of ages, into a menial position, and is an object, not of fear, but of patronizing sympathy and contemptuous indifference as to its threats, large talk, and puerile efforts at distinction or authority—to wit, Turkey, Spain, Greece, and Egypt.

The African's limited and child-like intellect, with strange notions of duty, right, and social privileges, tending strongly in the individual *as years advance* to assume a cruel and unfeeling regard for his neighbour, and only grows wise in cunning and cruelty as he gets older, is, through his intercourse with European blood, fast observing the evils he suffers through his service to the white man, and is ready to adopt measures for his own comforts and advantage, with an increasing desire with those advantages to maintain his own freedom; though, as far as the individual is concerned, he gradually is more cruel and cunning as his age advances.

And this increasing perception of liberty is the first spark for many ages, which has passed over the African nation, that gives a clue to his associating in many devices put forward by Europeans to turn his country into a large producing field for the good of all.

But it is going too far, as some would have it, to suppose that Africa was always equally debased as on its south-eastern and western coasts it is at present found; relics of civilized customs, and such as imply much knowledge and science, even still exist to mark that once a people of a superior position to those who now inhabit Western Africa lived on that continent.

Where went the Carthaginians, who in the days of Procopius were in a state of half-nudity, and occupying the confines of the old Carthaginian territory, when the conquering Saracens ran as swift spoilers through North Africa? One thing we know, namely, that the Vandals escaped to the mountains of Morocco; but where went these half-savage and enduring Carthaginians? Did any of them form into caravansiers, and, aided by camels, cross the Great Sahara and spread along the West Coast of Africa, and so relics of old customs pertaining to higher civilization found their way to Western and Central Africa? Such an origin is barely possible.

Leaving Africa, we get to the widespread, social, self-complaisant Asiatic, whose passive obedience and contentment have suffered a very material change by the blood of the restless and warlike Saracen. And now that she is pressed on the north by Russia, and on the south by Great Britain, and the stay of her contentment, her idolatry, is fast yielding to the ingress of Bible knowledge—without much Christian life, but with an intense interest to dive into the secret of European success and greatness—Asia on every side is adopting the principles of combination, in little and great affairs, to carry out the local schemes of any who are anxious to promote them.

Thus, it may be said, throughout the four great Continents and Australia, that the generalizing of *association* in little and great matters is gradually being stamped upon the present anthropological era. What its ending may be, or whether, as in the history of empires, about midway between it shall be on the wane, or give place to something more comprehensive and masterly, science has no sufficient basis in past history to speak with anything like decision; but it is well known, among a large number of the students of the Bible, that a great and widespread benevolence is confidently expected by a large number of Christian men, and that the principle of association will, ere this is consummated, be ushered in by one or two extremely extensive battle-fields, which, from the wide area from which the combatants will gather, would be totally impossible without considerable experience, and unless the minute ramifications of the general principles of association had taken an abiding place among the inhabitants of the entire civilized world.

Like epidemic eras, the anthropological eras bear a close analogy.

If, after the first two or three centuries of an epidemic, we have some new forms, or interlockings, or amalgamations, and towards the 450 or 500 years we have a leaning to decay, as in the Levant plague; or, like leprosy, changing its forms, and spreading wider at one time than another within given periods, so in the extension and rise of empires we have the same. Greece had existed as a power for nearly 300 years when it attained its highest glory; but little over a century had gone over its climax of glory, when it merged into a Macedonian monarchy, and, ere 500 years had passed over, its glory was fast fading and dying out; whilst Rome came from under her Samnite victories and Punic wars to be, within 400 years of her foundation, a consolidated power, which, for about 200 years, she had being making strenuous efforts. Her wide-spread dominion quickly followed suit, and, like the spread of leprosy in the twelfth century—from being but little known in South Europe, it passed its original bounds in a new epidemic era—so Rome attained in a new epidemic era, under Sylla, Marius, Pompey, Cæsar, and on to Diocletian, a supremacy in an anthropological sense, and a boundless sway over the civilized world, which mankind has never seen realized since. Then, for the last 150 years of this era, Roman power sank fast into the shades of increasing darkness.

From thence we see the Saracen arise, and Europe growing darker and darker; but in about 300 years to 350, the Saracen begins to wane—about 900 A.D.—and, by the time his era is ended, the blight of decay fast passes over his dominions. The Turk, the Tartar, and the hirelings in Egypt, slowly crushed and damped his ardour, and crippled his power on every side, and about 1237 to 1258 the Saracen fades away. And now Europe begins to progress slowly indeed, but surely, till about 300 years from the time of the new anthropological epidemic era of 1170 or thereabouts, and the dawning star of awakening intellect bursts into the sunlight of the fifteenth and sixteenth centuries, to go on with a slow progress of onward expansion, till it ends, in the early part of the nineteenth century, to take upon itself a wider and more general expanse throughout the entire globe; nursing, protecting, and sustaining this onward spread, through the agency of association, till consolidated strong enough, we may hope, in the strength and vigour of its own great power, it may stand upon foundations of merit and greatness sufficiently strong not to require the constant care of association-nursing, but be able to breathe the pure air of a more independent, mental, and social empire which shall be more abiding, and for the wider development of individual welfare and security to mankind.

Exception will be taken, in treating upon epidemics, that ethics and anthropology are here introduced; but as between the animal and vegetable kingdoms there are culminating points and strong points of analogy, as well as of contrast, so between the vital and moral force there is a mutual appropriation and welding of which it is well to give a passing notice.

If—through any peculiar telluric, geodic, and magnetic, or any other condition of the earth, or the adoption of particular habits, dietary, customs, or social virtues or vices, taken from a given centre, as Acre, we find, as ages roll on, a continued series of alternations, and wider and yet wider extensions and developments of social and dynastic empire, till it threatens to envelop the whole world, and to culminate, from its excellent geographical position, in making Acre its grand central depôt and great commercial market-place—it is inferred, from some unknown or known influence, that the organic functions *and nerve-enduring powers* of the great central cerebral masses are adapted within certain areas, and with races of men previously very differently fitted by customs, habits, and natural advantages, both of climate, soil, and geographical position, to be recipients of great metamorphoses, and to undergo very marked and unmistakable changes in the functions and powers of the brain— hence we recognize the brain and nervous system as the more immediate cause of successive changes of dynasties and empires, from a physico-vital point of view.

But with these changes in man's cerebral mass, *which is the great instrument of the mind*, it is well to observe that the pure physical power, or improved capacity and *endurance* of the brain and nerve centres, is turned, in an endless variety of forms, into a very widely destructive power by the *moral force* which rules and guides the physical power. For the physical power, which is the result of vital power or force shown in instinct or *identical* successive courses of action, under similar external conditions, is, for alien purposes, exceedingly circumscribed; whilst the *moral force* is wide-spread in its alien manifestations, and its aim in destruction is directed chiefly against accumulated property and cultivated lands, or all things which the hand of man aids in producing.

The watchful care and knowledge which enables man to plant and gather of the products of the earth, his neighbour, with like care and watchful knowledge, tries to destroy.

If, then, man's moral force is not, in so great an age of association and combination as is the present anthropological era, kept in check, and guided by some fixed and central ethical and comprehensive benevolent principle and system (and we have it already to hand in the New Testament), the result of great and ramifying association, with the ever-increasing means, by machinery and invention, of bringing that social principle into practical effect—from the world's past history, it is quite certain, if the moral system of an enlarged and practical benevolence does not gradually leaven the whole human race, that the vital force, through human devices of a destructive character, will expend itself in covering whole tracts of land with *thorns and briars and rank and useless vegetation*, in the place of the rich harvests of corn, and wine, and cattle, and all that is useful to both man and beast.

Hence, though the digression into anthropology, as given to us in history, may be considered somewhat foreign to a mere sketch of vital force, yet, when it is more carefully examined, a co-relation and balance between the moral and vital force meet in such a culminating point, and frequently in an antagonistic form, that the bare suggestion of their having a common field of action on this small planet of ours is scarcely out of place.

WE NOW COME TO THE SUPPOSED CAUSE, OR CAUSES, OF EPIDEMICS.

It will be perceived that neither earthquakes, volcanoes, comets, mildew, swarms of locusts, heavy continued rains, nor too long continued droughts, are admitted as the cause or causes of epidemic disease, in the broad sense of that word; but something more regular, constant, systematic, and always fraught with marked changes in the form of diseases, and in the type of those diseases, which are of a more constant and ever-present character—as inflammations and hæmorrhages, etc., as contrasted with new forms, like to Black death and cholera. And, with these changes in disease, the intellectual and moral condition of mankind have been closely associated, or at least an attempt has been made to show the *probability* of such a relationship.

It is well to remark that mildew, foggy weather, long frosts, long droughts, volcanoes, and earthquakes, and it may be in some very limited manner comets and eclipses, each have some effect in the promotion of epidemics, *but that promotion partakes more or less of an " endemic character."* In the case of volcanic eruptions and earthquakes, especially in those of the fourteenth and fifteenth centuries, certain emanations of a gaseous or material character may also have partaken of a poisonous character, and, to whatever extent those emanations extended, there specific diseases might have arisen as a direct effect of the diffusion of a poisoned atmosphere; yet, even under such circumstances, the spread of disease could scarcely be conceded to have much more than a widespread local or endemic influence. But in such a disease as the Levant plague or cholera, the regular advance and encircling diffusion over the globe, in the process of years, could scarcely be allowed to have its origin and diffusion as a sequence and effect of any particular emanation from any particular spot upon the earth's surface. Its march and invasion would forbid any such hypothesis, unless it be admitted that some particular emanation in itself cast out and diffused some special kind of vegetable or animal spore or fungoid diffusible germs, which as a permanent growth is highly improbable.

Even here we are met by the fact of such a disease as cholera, being an old and endemic form of disease, changing its character, and spreading more than the locust in every clime under the sun. How an eruption from the same centre, if such had been the case, could give new powers of diffusion and extension, centuries after, is as utterly inconceivable as that an eruption should have first originated in one part of the globe, and yet eruptions of apparently a like nature or character should have never had a similar effect in any other part of the globe.

Again, heat, either in excess or in defect, is a great factor of disease. The extremes between 32° and 120° Fahr. seem to be moderately well borne by vegetable life, and if we consider the teeming multitudes in India, China, and Central Africa, and of Central and South Europe, it appears to be equally well borne in animal life as represented in man, who, it must be observed, is a nude or clothed animal according to the exigencies of climate and customs, etc. But thirty degrees, above or below these points or ranges, is fraught with the most dangerous consequences to both animal and vegetable life; yet epidemic disease never appears to avail itself of these extremes for the end of encompassing with disease and death the multitude of victims which fall under its fatal grasp.

But it must be observed that during epidemic eras slow and gradual changes in the earth's temperature have taken place—but very much quicker, and of a far more marked character than ever could have been brought about within the historic age of man—by a slow decrease in the eccentricity of the earth's orbit, as maintained by the late Sir J. Herschel; which changes, from their speed, indicate something within the earth itself of a slow and gradual nature,

Epidemics.

but quicker and quite independent of the eccentricity of the earth's orbit.

For, howsoever we turn about, since the historic age of man, the difference with which the sun's rays have been received on the earth's surface, upon the score of the eccentricity of the earth's orbit, is merely fractional; and the diurnal amount from year to year on any particular spot of the earth is, from a practical point of view, the same now as it was in the days of Nimrod and Pharaoh I., if anybody knows on what day he ascended the throne in the days of Egyptian greatness.

Let us look at one or two of the indications of slow and gradual changes of temperature on the earth's surface within the historic period of man.

Job says that "my brethren have dealt deceitfully as a brook, and as the stream of brooks they pass away; which are blackish by reason of the ice, and wherein the snow is hid."*

Here the very description indicates ice in a pond, or covering a running stream; for though snow is white, when water by flowing has covered it and then frozen it in, the whiteness disappears, and a blackish look results. If such a passage as this is given, to illustrate the deceitfulness of character, to the hearers whom Job was reproaching, it cannot be doubted for one moment that both speaker and hearers were perfectly familiar with the fact which is here used as a simile for character.

To speak of the face of the deep, as in another passage of Job, being as stone and frozen; and a brook freezing snow in its course and covering it with ice in such a land as Idumea, or Edom, or Arabia Felix appears incredible; but from 1600 or 1500 B.C. this was a fact well known to the learned in that land, and familiar to a degree that

---

* Job vi. 15, 16; also see Job xxxviii. 30.

suffered it to be used as a matter of common illustration of subjects not otherwise so well understood.

Again, to speak of the coast of Greenland, now encompassed, ten, twenty, and thirty miles from shore, with a belt of ice, being in 982 A.D. perfectly free from ice, and in 1121 A.D. a bishopric appointed by Sigard, King of Norway, and in a short time having sixteen churches, one hundred and ninety hamlets, and two convents, appears equally incredible; but by the year 1408 ice so far increased round the coast, that Norway and Denmark ceased to count Greenland as a colony of any practical utility, and as utterly beyond the reach of maritime adventure.*

The Bosphorus was ice-worthy from land to land in 764, and 860 the Adriatic was frozen.

The Thames and the Baltic, and many rivers in England, were frozen between 1063 and 1683, especially from 1407 to 1683, and again in 1789 and 1814.

For the first four centuries of the Christian era the climate of England was not very inferior to that of Italy, and most of the fruits, especially the vine, which are peculiar to Southern Europe and Central Germany, were commonly grown in this country in the open air, and came to great perfection; but as a wine-producing country, for more than a thousand years, England has been as barren in this produce as the Isle of Skye or the coast of Finland.

From these and many other considerations, too numerous to be mentioned, it may be generally affirmed that within 3,000 to 4,000 years the more southern, and, proximally, the regions nearer the equator, have become hotter than in former times, and the more northern regions have become increasingly colder, with very marked and irregular exceptions, as in the years for this country of 1826 and 1868; but, be it understood, without any perceptible change in the

---

* See Scoresby's "Voyage to Greenland," 1823, introduction.

amount of the sun's rays falling on our planet from year to year during the whole of this period.

Under similar solar conditions our monsoons, siroccos, and trade winds, which, according to the solar theory, ought to come to the minute year by year, are far from obeying that law of uniformity; for so great is their deviation from this rule, that year by year the agriculturists and the tax-gatherers in India are looking for their advent with the greatest anxiety, as their frequent delays so materially affect the crops, which for want of sufficient rain within a given time determines their almost total failure, or contrariwise, an abundant harvest.

If it is not the sun's rays, nor yet the qualified intensity of them, through the spots or opaque masses on the sun's disc which are observable from time to time, what is the efficient cause of so much variation of temperature on the earth's surface, within given areas, which is capable of producing such an amount of irregularity? It will be said that the ever-varying conditions on the earth's surface, from the slow and gradual changes of depression and elevation in the masses resting on the surface, are the efficient causes.

But this will suffice. For, putting on one side the general and vague statements of older writers B.C., let us consider one or two of the remarks of authors since the time mercury was received as a standard of measure for temperature. Fahrenheit's being used in this country as the usual standard, it is here followed. Quoting from Haydn—

In 1796, Dec. 25, London, cold 16° below zero.
,, 1854, Jan. 3, ,, cold 4° below zero.
,, 1860, Dec. 25, ,, cold 18° to 15° below zero in several places in England.
,, 1810, Jan. 13. Moscow, mercury frozen hard.

Here four instances are given, and many might be added—as

from Dr. Short's work on the "Comparative History of the Increase and Decrease of Mankind in England, with a Meteorological Discourse," 1767, and many recent authorities—to show that changes in temperature are neither gradual nor equal, year by year, for the same locality. Here also consult the meteorological charts of Greenwich and Kew, or Paris and Berlin, when it will be found that thermometrical variations of 30 degrees are not very infrequent within 48 hours, for like times of the day or the night.

It is therefore clear that the mere surface of the earth, with its equal or nearly equal solar rays, cannot at all affect the equilibrium of temperature, by any sudden and violent changes within two or three days. Again, as all our storms and rain-falls are equally regulated by temperature, it follows that *no surface change* in electricity or magnetism can possibly accomplish such sudden and important variations.

Electric and magnetic variations, that are not sub-alluvial in their origin, must be accomplished by reciprocity of changes going on between lunar and solar rays, animal and vegetable organisms, and the general surface or stratified rock superficies of the earth; but the equal, gradual, and uniform nature of each, *cæteris paribus*, is such as to totally exclude the earth's surface as the great regulator of the sudden magnetic or electro-thermal variations now known to be perpetually occurring, and we must, as a sequence, admit *a central or internal source* as a regulator of the magnetic and thermo-electric conditions apparent on the surface of the earth.

These considerations lead to the inferences arrived at by Dr. Gilbert, 1600, and Halley at a somewhat later period—the former of whom judged that, towards the earth's centre, there was a large magnetic mass, and the latter that the earth consisted of an inner and outer magnetic shell, revolving one within the other, but at very nearly equal ratios of motion; but the difference of ratio was sufficient to account for the ever constant change or variation of the magnetic needle towards the Northern and Southern Poles.

In more modern times we understand that in both hemispheres, north and south, the poles have, in relation to the earth's axis, an eastern and western centre of magnetic attraction, or there are two northern magnetic poles and two southern.

The time in which each completes a revolution of 360° is variable. Dr. Barlow calculates the north-western pole to complete its revolution once in 850 years, and Haustein calculates the weaker pole as 860, and the stronger about double that time, or 1,740 in the northern hemisphere, and in the southern 1,307, and the stronger in 4,609 years. The observations and calculations, though accurate enough over a limited period of time, as those of the southern hemisphere, have been very recently made. But it is impossible, from the shortness of time over which these observations have run, to arrive at anything like a just estimate of the time required to complete any given revolution in the southern hemisphere, or eastern side of the northern hemisphere. Again, the declination or *secular variation of the compass* moves from east to west, and back again. In 1580 it was slightly to the east, in 1660 it was due north, and 1818 to 1823 it attained to its extreme westerly direction, and it has since been moving eastward.

To complete one oscillation, or complete extension of variation from extreme west to extreme east, it is calculated that a period of 320 years is required, and, of course, to regain the extreme west again would be another 320 years; which periods, added together, give the epidemic period here maintained as based upon historical data extending over 1,900 years, or from the time of Pompey's conquests in the east on to the present time, and not at all upon any magnetic theory or observations; since the historical data had been arrived at some years before the magnetic observations had been brought under notice.

The two works, from which the short abstract upon terrestrial magnetism is given, are, first, Sir W. Snow Harris's, F.R.S., "Rudimentary Magnetism," in Weale's Series, 1850; and, second, the article "Magnetism," in the second volume of "Natural Philosophy," published by the Society for the Promotion of Useful Knowledge. Much progress has been made, or theory suggested, since these works, but nothing that has materially shaken the outline broadly and yet briefly given in them.

It will be said that, between the time for the completion of an oscillation wave and its return, and a certain assumed historical data of 640 to 650 years, there is a lucky coincidence, and nothing more; but, as it is quite an undesigned coincidence, the historical data being assumed years before the oscillation wave was known, perhaps such a coincidence is worth while being considered.

As we have now arrived at a point where some general theory must be assumed as having some co-relation to vital physics and epidemics generally, a short summary of objections, and of things approved of in relation to some final theory or explanation, will be necessary.

1st: If the light and heat of the sun, aided by the reflected light of the moon, is equal year by year, and, with the exceptions of eclipses, spots on the sun, and comets, may be counted as actually invariable, it is legitimate to infer that they are not the cause or causes of the great variations of temperature, drought, excessive rain-falls, snow-storms, and frost, which occur year by year, or every few years, in all parts of the globe, and of such magnitude and variety as to be the source of great care and anxiety in our clothing, gardening, farming, and grazing operations; that suffering and even death are the direct results of the unequal distribution, at given seasons, of heat and moisture over limited areas of territory.

The famines of Persia and India will illustrate this statement, and in our own land the drought of 1868 might suffer feeble comparison in recent periods.

2nd: No changes on the superficies of the globe are of an extensive character, and at the same time sudden, but they are slow and gradual, unless we admit those from the interior which reach the surface, as earthquakes and volcanoes; but it is no infrequent thing, within 48 hours, to experience a change in temperature of 30 degrees Fahr. Such sudden changes are impossible to result from chemical decomposition, when they range over miles of territory, and much less can the regular solar heat vary so suddenly. Hence the changes arising from solar heat or chemical decomposition are totally insufficient to account for those frequent and sudden changes of temperature and moisture so commonly experienced, or for those changes of a longer duration and of a more subtle nature, called epidemic periods, which so materially affect the vital conditions of particular districts, and of large portions of the globe itself.

3rd: Force, in some way or other, holds and binds this earth together—call it attraction of cohesion, aggregation, or attraction of gravitation. It is force of some kind which holds this earth together. But to be force, as already mentioned in the chapter upon vital physics, *antagonism* is essential.

And the earth itself has within itself the antagonizing forces of repulsion and attraction, but in such very different degrees, and manifested in such a variety of forms, that the whole earth is in unceasing action throughout, in the forms of magnetism, diamagnetism, internal heat, and fire, with the outcome of constant processes of elevation and depression; whilst, by the very nature of its rock superstructure or stratification, an enormous amount of dry, moist, or wet galvanic action is going on, which in less than 50,000 years bids fair to destroy, by its constant disintegrating effect, every petrified remain contained in our stratified rocks; and the present state in which they are found too surely marks the slow process of disintegration, or decay, to which they have been subjected; whilst through some chemical process, carried on by electro-magnetism and galvanism chiefly, our forest of Coniferæ have undergone a half crystalline and stony change in the form of coal beds. Added to these, such constant changes are going on in the magnetic dip, and in the position of the magnetic poles, that it is impossible to deny, that within herself, the earth has constant and ceaseless work going on, and by forces in perfect unison, and in perfect harmony with those forces which bind the earth, the sun, and all the planets to each other.

Hence it is inferred that, as certain actions and forces, between this earth and the sun, are essential for the due display of vital phenomena on the surface of the globe in the form of animals and vegetables, so for the preservation and due integrity of vital forms existing in the two kingdoms those forces which are within the earth, also, have a certain and constant effect upon all things moving and growing upon its surface; and that, from the changes and balancing of forces within, the vital organisms receive energy, or are retarded in those functions and forms of manifestation which they exhibit on its variable and ever-changing surface.

But, from the nature of epidemic and many other phenomena, it is assumed that there are regular and consecutive changes constantly going on within the earth, but not of that equal and invariable character observed in the motions of the planetary bodies; and that, at some time or other, *some violence has occurred* to its internal constitution or mechanism, which has thrown out of order that uniformity of sequences, so much observed in our seasons, and our diseases, yet not to that degree in which general laws cannot be distinctly traced and plainly indicated.

That violence, or a succession of violent shocks, upon a large and gigantic scale, have occurred to this globe, geology has long since recognized. From the views already given in relation to the earth itself having, from the constant relation of forces within itself, a true and certain effect on vital manifestations on its surface, it will be perceived that no changes, of any great magnitude within, can occur without, in some measure, the vital integrity of objects on its surface being more or less affected; but changes far greater, and of a more powerful nature, than anything merely going five to ten miles below the surface.

The great question, therefore, is this—Is stratification the result of slow and gradual changes, running over a long series of years and ages, till we gradually arrive at the expression of an epoch, may-be, of 1,000,000 years, or twice or three times that time, to be succeeded by another epoch of equal or longer duration? Or is rock stratification owing to some mighty and much more comprehensive catastrophe, and is the present order of things simply the outcomes of all the good and evil which that mighty change has left to mark its completeness and its existence?

To this latter view the writer is disposed to lean, no matter what the enormous difficulties are which appear to stand opposed to it; and in 1864 he wrote a short tract anonymously, entitled " Doubts Relative to the Epochal and Detrital Theory, by a Near Kinsman of Thomas

Didymus," which tract has a very direct bearing upon the whole subject.

The entire evidence of a succession of changes and convulsions of the earth, or only one mighty and exhaustive one, turns upon the subject of rock formation *from detritus*, and this paper will conclude the present short treatise on vital force, and is, as it were, a kind of corollary to the entire subject of epidemics.

DOUBTS RELATIVE TO THE EPOCHAL AND DETRITAL THEORY OF GEOLOGY.

As epochs or eras in geological language are not accepted as representing the exact opinions held by most geologists at the present time, and any attempt to disprove the epochal or detrital theory of geology, based upon the notion that *every stratum* has been formed at a particular period of time, and that during *the same period of time* none other has been in process of formation at some distant portion of the globe, is totally unnecessary; since most geologists are prepared to admit that upon distant areas of the earth's surface, in one part a given stratum may be completed, whilst at another part a new stratum is in process of formation. Thus free scope is left for the application of the developmental theory of Darwin, upon the basis of natural selection, without the necessity of supposing each new species requiring a special act of creative power for its manifestation, but only *an improved condition* favourable for its development.

But whilst the distinction of epochs or eras is rejected as being totally inapplicable to any particular formation, as regards the period of deposition *in all parts of the earth at one and the same time*, yet, on the other hand, it is maintained that, in any given area or locality on the earth's surface, every stratum there found super-imposed upon another, is as distinct and separate in relation to time as the several strata therein contained are distinct from each other.

Hence any given locality, from the strata it contains, correctly expresses according to the detrital theory, by the order of super-position, the relative periods or epochs at which each individual stratum therein contained was formed; and, also, from the depth and contents of each stratum, the length of period required for its formation is surmised.

It is the purport of this paper to question the validity of DETRITUS, as being a sufficient means of explaining the formation of strata, and, therefore, of the value of *Stratification* as a means of determining the age of the earth, or how far we are justified in viewing it as a chronological chart.

The original crust of the earth, whether regarded as of aqueous or igneous origin, is admitted on all hands to have been GRANITE, and is thus described by a well-known author:—" The unstratified or igneous rocks occur in no regular succession, but appear amidst the stratified without order or arrangement, heaving them out of their original horizontal positions, breaking through them in volcanic masses, and sometimes over-running them after the manner of liquid lava. From these circumstances they are in general better known by their mineral composition than by their order of occurrence. Still it may be convenient to divide them into three great classes—*granitic*, *trappean*, and *volcanic;* granitic being the basis of all known rocks, and occurring along with primary and transition strata; the trappean of a darker and less crystalline structure than the granitic, and occurring along with the secondary and tertiary rocks; and the volcanic, still less crystalline and compact, and of comparatively recent origin, or still in process of formation."

It is, then, from this primary rock, *granite*, that we ought to seek for the elements found in subsequent rock formations, which take their origin from the granitic series; oceanic currents, winds, rains, floods, frosts, thaws, heat, rivers, and seas, all aiding to disintegrate and drift from one point to another fine particles of dust, sand, and detritus generally, accumulated from the attrition of granite in the first instance, and afterwards from rocks formed out of such attrition, and exposed to like disintegrating agents as first acted upon the granite, each assisting by their own materials in the formation of subsequent strata. But in the secondary and tertiary formations, vegetable and animal organization largely contribute to fix the character and structure of considerable portions of particular strata.

Such being a general outline of the formation of strata, and the source from whence they proceed, it follows that from this primary rock, granite, we ought to seek for the elements which make up the more recent ones, *which are said to originate from it*. But if elements are found in these rocks which granite does not contain, or if elements are contained in very great abundance in the strata above granite, and in granite they are found in very small proportions, it is evident that they are not derived from the mechanical disintegration of that rock, aided by the accumulation of depositions in the form of organic remains.

From the foregoing general observations it will not be deemed irrelevant if the attention is now directed to the ORIGIN OF LIME, as a constituent of most stratified and transitionary rocks, in some occupying a limited range in their constituent parts, and in others by far the greater bulk; whilst at all times it is very limited in granite.

The most important step in such an inquiry is to ascertain, as near as may be, the chemical composition of granite;* and to this end a table (No. 1 Table) is supplied, chiefly gathered from a FEW of the analyses of Bischof, and added to those of Brande and Page. In this table many minerals known to exist in granite, but which are only occasionally found, such as orthite, apatite, &c., are designedly omitted, as also is garnet (which occurs more frequently in eruptive rocks than in granite, and which, moreover, in its chemical composition in relation to lime is very variable, lime being sometimes altogether absent), because the quantity of lime derivable from such sources is too inconsiderable to deserve attention.

TABLE I.

		Silicon.	Alumina.	Lime.	Potash.	Soda.	Magnesia.	Water.	Iron.	
Brande's Dictionary of Arts &c.	Felspar	67	19	..	14	..	..	..	..	Often found among the eruptive rocks.
	Mica	46	14	..	10	..	10	..	20	
	Quartz	100	..	..	..	..	..	..	..	
	Hornblende	59	..	14	..	..	20	..	7	
	Steatite	59	..	..	..	..	32	7	2	
	Serpentine	42	..	..	..	..	40	15	3	
	Chlorite	52	10	..	7	..	12	6	13	
Dd.Pge.	Hornblende	46to6c	..	7 to 14	..	..	14to28	..	..	
Bischof. Mica. Felspar.	Orthoclase	65·52	17·61	0·94	12·98	1·70	..	..	0·80	
	Oligoclase	63·70	23·95	2·05	1·20	8·11	trace.	..	..	
	Albite	69·00	19·43	0·20	..	11·47	..	..	..	
	Adular	Same as Orthoclase and Albite.			equal parts of Potash and Soda combined.		..	..	..	
	Magnesia Mica	44·63	16·48	..	9·57		19·06	1·25	11·52	Eruptive rocks.
	Potash Mica	48·00	34·24	..	8·75	..	0·50	..	4·50	
	Labradorite	52·52	30·03	12·58	..	4·51	0·19	..	1·7	
Dd.Pge.	Augite	53	5	19	..	..	15	..	6	

* In examining the analyses of granite, a careful distinction ought to be made between the granitic and granatoid rocks.

TABLE II.*

Tertiary or Cainozoic.		Mineral accumulations of historic period, Pleistocene, Pleiocene, Miocene, Eocene.
Secondary or Mesozoic.	Cretaceous.	Chalk of Maestricht and Denmark, Ordinary chalk with and without flints, Upper green sand, Gault, Shanklin sands, Vectin Neocomian or lower green sand, Wealden clay, Hastings sands, Purbeck series.
	Jurassic or Oolitic.	Portland oolite or limestone, Portland sands, Kimmeridge clay, Coral rag with its grits, Oxford clay with Kelloway's rock, Cornbrash, Forest marble and Bath oolite, Fullers' earth, clay, and limestone, Inferior oolite and its sands, Lias upper and lower with its intermediate marlstone.
	Triassic.	Variegated marbles, Muschelkalk, Red sandstone, grès bigarré, bunter sandstein.
Primary or Palæozoic.	Permian.	Zechstein, dolomitic, and magnesian limestone, Lower new red, conglomerate and sandstones, Coal measures.
	Carboniferous Limstne.	Carboniferous and mountain limestone, with its coal, sandstone, and shale in some districts. Carboniferous slates and yellow sandstones.
	Devonian.	Modifications of old red sandstone.
	Silurian.	Upper Ludlow rocks, Wenlock shale and limestone, Woolhope limestone, Middle Caradoc sandstone and conglomerate, Lower Landeilo, Bala, and Snowdon beds.
	Cambrian.	Barmouth sandstone, Penrhyn slates, Longmynd rocks, and various rocks below the Silurian.
Hypozoic.	Mica Schist.	Beds of mica schist, consisting of quartz and mica with or without felspar or garnets, chloritic schist, talc schist, quartz rock, clay slate, limited beds of iron ore.
	Gneiss.	Beds of gneiss, consisting of laminæ of quartz, felspar, and mica; beds of mica schist, quartz rock, limestone, hornblende schist.
Primitive or Igneous.		Syenite, Porphyry, Basalt, Porphyritic granite, } The relative position and age of these rocks is more or less uncertain, though it seems probable that they may stand in the order here assigned to them.
		Granite, { This system of igneous rocks descends to an undefined depth, and is assumed to rest upon the internal liquid nucleus of the globe.

* From Lardner's "Popular Geology."

In examining the rocks formed from granite (including their organic remains), there are none which require a more careful consideration than the gneiss system, the same being first in order after granite, but in chemical composition containing an excess of *lime* as compared with that rock. In this system hornblende schist occupies a prominent place, in conjunction with talc, chlorite, and mica schists, amongst which latter *magnesia* takes a more prominent place than lime.

The texture and appearance of the metamorphic rocks closely resemble that of granite, and the extreme comminution of their crystals closely approximates the primary plutonic rocks. They *appear* to consist of little else than crystals of granite, worn and comminuted, and thrown together into strata, or thinly laminated beds of granite *débris* deposited from water, which had worn the granite rock in one part of the ocean, and by currents conveyed it to another.

By observing Table I., the per-centage of lime and magnesia in hornblende exceeds that in mica or felspar very considerably; and hornblende is a rock very generally distributed in the granite series, though more superficial and more limited in quantity than either felspar, quartz, or mica. Hence hornblende is naturally looked upon as the source of lime and magnesia in the metamorphic rocks, if these rocks are formed MERELY by the disintegration of the granite. But in the gneiss system hornblende crystals occur in such great abundance as hornblende schists, that it must be inferred that this schist, in the gneiss, represents fully and ENTIRELY all the hornblende derivable from the disintegration of granite.

Now, in the gneiss system, dolomite (composed of equal proportions of lime and magnesia), is occasionally found, and the talc schist in the same system contains as much as

36 per cent. of magnesia. It, therefore, cannot be supposed that the hornblende in the granite, which is pre-occupied in forming the hornblende schist of the gneiss system, can also suffice to yield in addition as much lime and magnesia as occur in the dolomite and talc schist of the same system; and still less that it should, beyond that, suffice to account for the lime occurring as carbonate of lime in the primary limestone of the gneiss system.

Again, even supposing that the hornblende schist has had both magnesia and lime abstracted from it, and in places here and there it be found to be considerably freed from them, yet the amount of hornblende or syenite is so inconsiderable as compared with quartz, mica, and felspar, that it is totally insufficient to supply the amount of crystalline limestone and of lime contained in dolomite found in the metamorphic rocks; and the mere fractional amount contained in felspar is scarcely worth a serious consideration in accounting for so large an amount of lime as is found in the primary limestone and dolomite belonging to the gneiss system.

If then lime, as contained in granite, is more than exhausted in supplying the first series of the *transitionary rocks*, and that alkaline earth is there taking precedency over its representative in granite, what can be said of lime in the strata above them? Passing over the Skiddaw system of rocks, in which hornblende and chiastolite slate are largely supplied with magnesia rather than lime, we come to the Cambrian and Silurian systems. In these rocks, whether the Bala or the Llandeilo, the Plinlimmon or the Wenlock rocks be examined, there we find LIME in considerable abundance, and among the alkaline earths taking the foremost position. From the silurian onwards, through the Devonian, carboniferous, new red sandstone, on to the chalk, *lime* is found amongst the alkaline earths in

Detrital Theory of Geology. 305

by far the greatest abundance, magnesia the next in frequency, and potash the least; neither is this order of amount or frequency changed when the tertiary formations are carefully considered.

For if the chief elements or earths were arranged in the *order of frequency or quantity* in the granite, and then the sedimentary rocks, they would stand pretty nearly in the following order: in granite—silicon, alumina, potash, magnesia, iron, soda, and lime, would represent the relative amount of the component parts, commencing with silicon as the most abundant, and lime as the least; whilst in the sedimentary rocks the arrangement of the constituent parts being made in the order of frequency or amount, the following would obtain: silicon most abundant, alumina and lime nearly equal, and magnesia, iron, soda, and potash much less abundant.

After a careful consideration, by the data supplied, of the component parts of the sedimentary and granitic rocks, though there is no certain measure, it is probably not far from the truth to estimate lime in granite as about 3 per cent., and in the sedimentary 16 or 17 per cent. If such, then, be an approximation to the actual relation of lime as contained in the two kinds of rocks—plutonic and sedimentary—whence is it that such a contrast in the quantity of LIME should exist, when one is derived in part or entirely from the other? In the mode in which the matter is here put, it will be said, nothing can be admitted which does not involve the absurdity, that the effect is greater than the cause. Therefore, for the better appreciation of this important subject, one or two of the usually supposed sources of LIME will be examined.

As, before there were rivers, the mighty ocean appears to have washed over or rested upon ALL rocks, so in its waters will be found all the soluble salts not reduced to an insoluble rocky precipitate, and what lime at that time was not a component part of granite, etc., was a component part of the oceanic waters. Again, taking the wide expanse of the ocean as it now exists, and from its intensely salt flavour towards the Equator, and its slightly salt flavour in the Northern Ocean, especially in the Arctic regions, the waters between the Mediterranean and the British Channel may be considered as a fair average. If, then, the two analyses be taken, the one by Dr. Schweitzer for the British Channel, and the other by M. Laurens for the Mediterranean, the following results are obtained :—

TABLE III.*

	English Channel.	Mediterranean.
Water	964·74372	959·26
Chloride of Sodium	27·05948	27·22
,, Potassium	0·76552	0·01
,, Magnesium	3·66658	6·14
Bromide of Magnesium	0·02929	...
Sulphate of Magnesia	2·29578	7·02
,, Lime	1·40662	0·15
Carbonate of Lime	0·03301	0·20 c Mg.
	1000·00000	1000·00

With the results set forth in Table III., let it be granted that if all the lime could, at one given instant, be precipitated from the ocean, even then, the deposit would not form a seam or stratum of two yards in perpendicular depth; which would be a mere fraction compared with the actual amount of lime present in the successive sedimentary rocks. Its very sparing solubility, compared with soda and potash, *except as a chloride of calcium*, excludes the supposition of its having entered largely into the composition of sea water. And if it had been contained as a chloride, sulphuric acid is the only acid found in sea water capable of precipitating it,

---

* Graham's "Elements of Chemistry," 1st edit., p. 266.

when it would have appeared as insoluble gypsum—a rock found here and there in the sedimentary series, especially in the triassic and tertiary systems, but nowhere to any great extent. On the other hand, if it be imagined that *heat* has expelled the chlorine, and so a large sediment of lime has been precipitated, still, setting aside every difficulty as to where the chlorine went, and how it was again taken up to form a chloride of sodium, etc., it may be briefly stated that such a process would only assist to explain the formation of ONE STRATUM or layer, and would leave the rest to chance.

Hence the conclusion of Bischof may be adopted without hesitation, that—" The assumption that sea water contained a larger quantity of carbonate of lime at the period of the formation of the great limestone strata from the transition limestone to the chalk, and that the increase of limestone formations during this period was a consequence of the decrease of this carbonate in sea water, is contradicted by the circumstance that it would then have been impossible that a solution should have been left which is so far from saturation as the sea water of the present time ; *for all precipitations which result from evaporation of solutions leave a saturated mother liquor.*

" It is, therefore, evident that in every point of view the assumption that our great limestone strata, from the grauwacke limestone to the chalk, have resulted from the evaporation of sea water, is altogether unfounded."*

It being granted that the LIME in granite, superadded to that of the lime in sea water, is insufficient to account for the amount found in the sedimentary rocks, as now known, the important point still remains to be answered—Whether the calcareous casts or exuviæ of the universally diffused

---

* " Chemical and Physical Geology," by Gustav. Bischof. Vol. I., page 178.

families of the foraminifera and polypifera have not been the efficient factors of our *lime strata*? The bare fact that much of our lime, especially in the oolite and cretaceous systems, is nothing else than organic exuviæ, is too patent to admit of a moment's doubt as to its correctness; but the difficulty does not rest with the admission of such facts, since the original doubt still presents itself—Whence do these polypi obtain their lime? The answer to this question usually is, that however small the quantity of lime sea water contains in a given measure, yet such is the industry of the madrepores, astræas, the millepores and seriatopores, that islands are raised by them in the midst of the ocean, and a continent is being girt by their increasing labours. Granted that it is so, yet the BED of the ocean in many parts, and the washing of its endless SHORES, yield a constant fresh supply of lime to those waters from which the lime has been abstracted by the industrious zoophytes. The founding and extending of coral-reef building has nothing to do with the small amount of carbonate of lime *held in solution* in sea water *at any one given time*, but whether, when that is exhausted by the reef-building polypi, there is placed within the limits of the ocean's expanse lime sufficient to replace that abstracted from the waters, and so the unceasing *demands* of the polypi receive an unceasing *supply*.

For if there is not sufficient lime, then, where that limit defines itself, there their work must cease—since zoophytes are no more alchymists than the higher order Bimania—and the power to convert silicon or alumina into lime, where the latter element (or compound) ceases to exist, would be necessary to enable them to carry on their work; for they cannot create matter or lime, they can but apply it when present. Therefore, the organic source of lime is bounded by the same limit as that of pure sedimentary rock—namely, *the quantity deposited as exuviæ is limited by the sources or means of supply.* If, furthermore, it is maintained, that the ocean has within it springs of carbonate of lime, yet those springs cannot have their source where granite is the bed; for what it contains in a very small proportion, it cannot supply in a very large one. Hence, from whatever side it is viewed, the *lime* contained in coral-reef, etc., cannot have its source in or from granite.

Another source of lime must not remain unnoticed—namely, the eruptive rocks that are not granite, for M. de Beaumont has shown that these latter rocks have protruded themselves through strata even later than the secondary, upheaving, dislocating, and displacing their previously horizontal beds or plains.

As far as the eruptive rocks stand related to lime and stratification generally, there is much difficulty in deciphering their origin and effects, and they are closely connected with the internal heat of the earth. The seething waves of liquid fire, which, on approaching the crater of an active volcano, can be distinctly heard, and which, after being long pent up, pour forth their fiery streams as a devastating scourge upon all around, bespeak the potency of the element at work, but, like their own clouded light, leave us in ignorance as to the extent of the hidden caldron, or its location in the bowels of the earth.

Though the heat beneath our feet increases by regular increments, after a certain depth from the surface—say, in temperate regions, from 60 to 90 feet, and by calculation the earth is proved to be in a state of fusion at a comparatively short distance from the surface, yet the mean density of the earth, as proved by the oscillations of the pendulum, forbids the notion that the interior can be lighter than the crust, or that the liquid lava poured out from a volcano can have the same specific gravity within as on the surface, if it be liquid at one fourth the semi-diameter of the earth, and that liquidity be produced by a degree of heat of far greater intensity than can possibly be produced on the surface.

Under such conditions of heat, and allowing with Dr. Young that "at the earth's centre, steel would be compressed into one-fourth and stone into one-eighth of its bulk,"[*] yet, as the heat is still increasing as the centre is neared, it is difficult to conceive how molecular attraction could resist the expansive power of heat, and that any other than a perfect fluid of a homogeneous nature, and of no very great specific gravity, occupies a very considerable portion of the earth's interior; the researches of Airy and Hopkins have gone far to determine such a condition.

In contrast with this conclusion, and also with that now very generally adopted, of the earth having once been a fluid or molten mass moving in its orbit as a fiery meteor, the two following facts relative to the earth's *surface* appear to be in direct opposition.

1st: From whatever point of view the internal heat of the earth is measured, the external area of the globe cannot be more contracted than it has been for ages past, for the amount of *unconformity* in the *sedimentary* rocks is less in proportion, so far as observation has yet extended, to the amount of *disruption* of granite, or of irruptive rocks, since such sedimentary rocks have been formed; which circumstance bespeaks rather an increase than a decrease of the earth's superficial area. Moreover, the general elevation of the sedimentary rocks from the time of the metamorphic rocks until now, as shown by the increase of LAND plants and animals, especially AFTER the formation of the Cumbrian, Cambrian, and Devonian systems, all tend to the same conclusion. For during their formation the almost total absence of land organic remains, and, in some parts, the great abundance of *marine remains*, attest the *univer-*

---

[*] Lyell's " Principles of Geology," 8th edit., page 515.

sality of sea distribution, and the *remarkably slight elevation of the land;* whilst, after these formations, land remains become more extensive and abundant.

If, then, it is admitted by geologists that the aqueous and atmospheric oceans are and have been the same in all ages, for the increased elevation of land above the sea, it must be admitted that the *earth's area* has, in process of ages, *expanded* rather than contracted, and that the gradual gaining of the land over the sea *is owing to an increase of expansive power or heat, and not to a decrease; for all forms of matter, save that in the form of water,* CONTRACT *when they have passed from a fluid into a solid condition.* The flues of the earth, in the form of active volcanoes, are but relics of this elevation, and are so many safety-valves to moderate and equalize the relation between land and sea, and to check too sudden or unequal elevations between the bed of the ocean and the land. At least, from known data, such an interpretation is perfectly valid.

Cracks and fissures from the cooling and contracting of the molten mass, allowing the protrusion of more deeply-seated porphyra, syenite, etc., etc., would only abstract heat and matter from the interior to be placed upon the surface, and would thereby occasion depression and increased submergence of land *elsewhere*, and could not in any wise lead to the gradual gain of the land over the sea, but, so far as the *heat* is concerned, to *positive loss* by radiation.

The known *decrease* of the *solar heat* upon the earth's surface is perfectly compatible with the *increase* of the *earth's area*, since by it *the surface of radiation is increased;* and if the atmosphere through which the sun's rays penetrate is the same, then, the superficial area being increased, rarefaction will be increased, and positive heat will be rendered latent, and so the mean temperature of the earth will be slowly diminished; it being assumed, at the same time, that the atmosphere is the same now as then, and that its amount of aqueous vapour be constant, and so the effect of the chemical or calorific rays of the sun were brought into action then as at the present time.

The slow decrease in the eccentricity of the earth's orbit ought also to be borne in mind, as a further source of diminished solar heat.*

Should it be said, on the one hand, in opposition to the aforesaid conditions of increased superficial area and diminished heat, that such increased area is merely fractional in relation to the diameter of the earth, it ought, on the other hand, to be remembered, that the diminution of solar heat, as determined by positive data, is merely fractional, though it is well proved.

2ndly: Let it be granted that the space included between the two circles, " out, out, out," and " in, in, in," represent the solid crust of the earth, as set forth in the accompanying diagram; and let A, B, C, be an equilateral triangle, whose base shall extend from B to C, and whose apex shall be A. If, then, force be acting equally in all directions, as from the inner circle of the earth's crust, " in, in, in," which inner circle forms the outer boundary of a great caldron, and at the point A the crust shall yield, then the lines F F F will represent the direction of force from the circumference to the sides of the triangle, and the equilateral triangle A B C will give the widest base conceivable, in the straight line B C, in which it is possible that force, effecting a rupture at the point A, through the earth's crust to the point P, can attain.

After making all due allowance for the thickness of the sides of the mass elevated upon the earth's surface from such a base to the height of five to ten miles, *it is impossible*

---

* *Vide* Sir J. Herschel's " Discourse on the Study of Natural Philosophy," 1831, p. 147.

PLATE III.

Page 312.

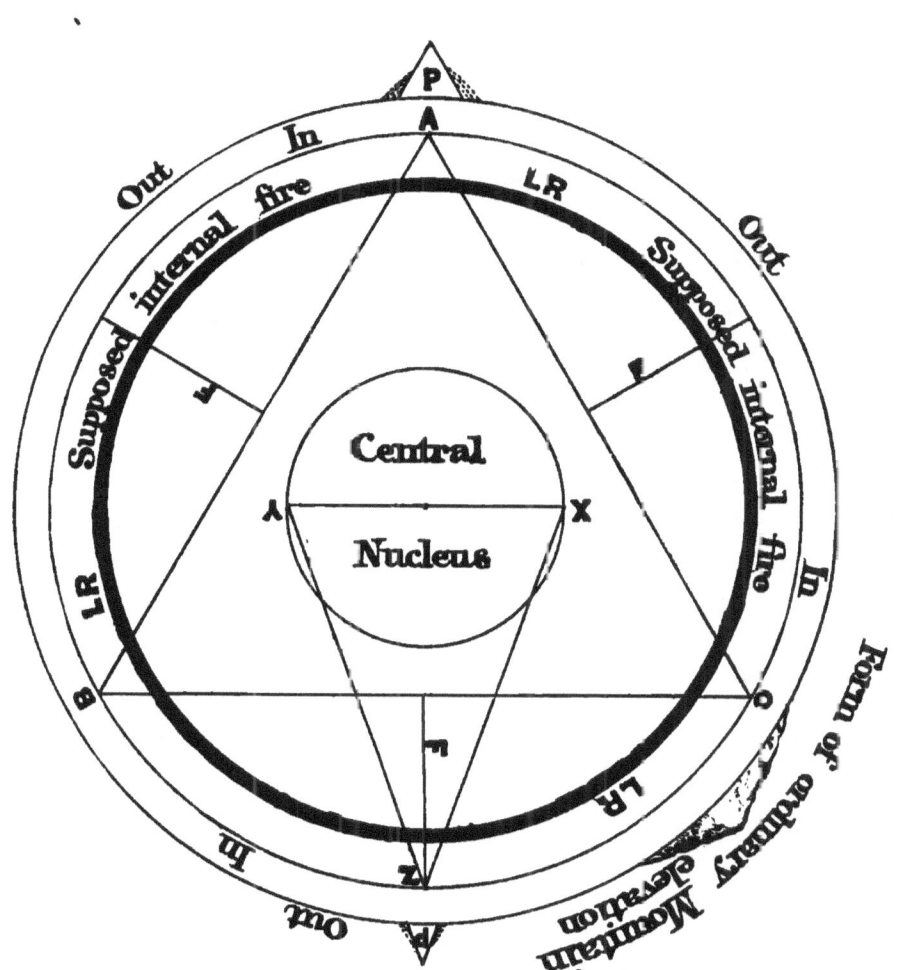

to conceive the mountain mass to have a base much wider than that belonging to the base of an equilateral triangle.

But, on the other hand, grant that the earth has a *central nucleus*, and that the heat has not rendered fluid all the matter from the inner circle to the centre, then let X Y be the base of the isosceles triangle X Y Z, and let Z be the apex.

As the central nucleus, being solid, must be the fixed resisting medium for all force gendered by heat acting towards the earth's surface, then, as the base from which that force acts cannot be wider than the diameter of the central nucleus, however large or small that nucleus may be, it follows that, for the crust to yield opposite to the apex of a triangle so formed, as at the point P, opposite Z, the elevation on the surface formed by force so acting, must have a base *far narrower than the length of either of the other two sides* leading to the apex.

Hence, if force act either from a fixed central body, or from the sides of a large enclosed caldron, the elevation on the surface arising from it, *and only acting at one point*, so as to form a mountain or peak, will have the sides *as long or longer* than the *base*, as the case may be, *but under no circumstances will the base be materially wider*, the discrepancy being little more than fractional.

In contrast with this induction, stands the formidable phalanx of all the mountain systems in the world, of every hill, and, almost, hillock, found on land; for at the *base* of each, whether it be of the Himalayan range, the Andes, the Alpine, or any other mountain range, let a bore or tunnel be cut through, and in many the base would be quite five times longer than the height taken at its highest peak; and there are few which would not have a base three times longer, or wider, than the height; while, in a mountain formed upon the basis of an equilateral triangle, the base would not be nearly so much as twice as long as the height, nor more than equal to the sides, but considerably under that proportion, if raised from a central nucleus.

To obtain a *base* sufficiently broad to meet the case *as it is actually found*, *a line of resistance* much more superficial than from a supposed central nucleus, or from the sides of an enclosed sphere, must be assumed, in order that *the base on the surface* may be formed by force acting *from a surface below* sufficiently extensive to secure the base known to exist on the surface above.

Let it, therefore, be granted that the circle "in, in, in," contains a smaller circle within it, L R, L R, L R. The same circle shall be an *unyielding circle*, and be called the "*line of resistance.*" If, then, between the two circles there is an expansive force, as fire, acting upon the earth's crust (which crust may be supposed to be less deep than represented in the diagram), then, according to the area of resistance occupied along the line or circle of resistance L R, L R, L R, so would be the extent of the base below to the surface above belonging to any particular mountain range or peak, the requisite amount of force being *a priori* assumed as a necessary element in the postulate.

The extent and degree of elevation of our table lands, steppes, and mountain ranges, ought, in time, by careful calculation, to lead to a very close approximation as to the depth from the earth's surface of *the line of resistance* underlying the heat.

Against this view, it will be said that too great an amount of solid matter would be admitted to account for the mean density of the earth. Granted, if it were necessary to admit the whole to be solid beneath the line of resistance; but a fluid so slightly compressible as water might be admitted as occupying a portion of the space, especially if the earth, in conformity with all organic bodies, has within itself a fixed internal structure; and if not, yet the form, proportion, and extent of the great mountain chains, or backbones of the earth, our table-lands and steppes, ought to demand that the formation of their physical construction be explained independent of all ulterior facts that may hinge upon their elucidation—since careful induction, *from well-known data*, can only lead to greater light and higher truths.

From the foregoing reasons—namely, 1st, the gradual increase of land over sea, which has been going on through successive orders of strata; and 2ndly, from the nature of conformity in our mountain systems and table-lands, etc.—it is inferred that on the earth's surface we have nothing lower in order of stratification than granite; for if a lower form of rock existed capable of transfusion between the granite (and not veins and masses of particular forms of melted granite), surely the conditions requisite for its transfusion have been abundantly supplied both by the gradual extension of the solid area of the globe, and also by the comparatively superficial depth at which the heat is located, and which would have forced other matter above, in a manner sufficiently plain so as to lead to no doubt as to its lower origin, both from its own character and the extent of surface it would occupy at different districts over the globe; but granite is evidently sufficiently thick and extensive under all circumstances to fill up, by its own melted substance, all cracks, dykes, and fissures that may occur from the loss of continuity in its own substance, as is continually seen in the vitrified masses of granite running for miles in particular directions, and materially interfering with the quarrying of granite in the numerous Tors in this and other countries, and known as "horse tooth."

Hence the conclusion drawn is, that granite is the true primary rock, and that trachyte, basalt, trap, or greenstone, etc., etc., are but melted forms of either granite or some one or more of the metamorphic or secondary rocks, and that the eruptive rocks are in origin above and not below granite, which is the generally received opinion.

But if it be granted that the eruptive rocks are BELOW granite, yet their extent is too limited, and the TIME of their disruption in the order of stratification too recent, to admit them as agents that can have supplied much of the material for lime and magnesian limestone found in the sedimentary rocks, from the gneiss system to those of the cretaceous or tertiary formations.

Hence the eruptive rocks, as agents in supplying lime to the sedimentary ones, are insufficient, whilst their origin is hidden in much obscurity.

Having examined the origin of *lime* as derived from granite upon the detrital theory, and found it wanting, it is not unfair to inquire how the case stands as regards the POTASH in both felspar and mica. As both these components of granite contain much more potash than lime—say five of potash to one of lime—we ought, by the process of mechanical disintegration, to have for every foot in thickness in lime, about five feet of potash. And if, in addition, the laws of solubility are to be taken into account, then, as potash and soda are much more soluble than lime or magnesia, the excess of potash over lime ought to be greater still; and instead of being 1 to 5, it ought to be nearer 1 to 8. But what is the fact? In the whole geological series of stratification, there is not one single potash stratum, only given beds, like to common salt.

Now, water disintegrates felspar, and with it the contained potash, as most engineers connected with water-works will decidedly affirm. And upon this subject Graham thus speaks:—" A comparison of compact and disintegrated felspar shows that by the solvent action of water, the latter has been deprived of half its silica, and above three-fourths of the potash."* Hence the non-appearance of potash as the

---

* Graham's " Elements of Chemistry," 1st edition, page 522.

sole or chief ingredient of any stratum formed from granite detritus cannot be attributed to the resistance which that rock is capable of offering, by reason of its compactness, to the solvent action of the water on the potash contained in it.

But it will be said that sea water, fresh water, land plants, and all alluvial soils, especially those of India, and soils not far distant from mountain ranges of granite formation, have potash; and though unaggregated into distinct strata, yet, as it pervades all strata, or NEARLY so, its amount must be very great, and in a measure proportional to its excess over lime in granite. If such be the inference, then, how does it happen that in all soils and strata where potash is found, there lime is found, and almost invariably at a higher rate of per-centage than potash in the same soils? In addition to which facts, *lime is found aggregated into distinct and special strata of its own elemental composition.*

To sum the whole matter up. Nothing but difficulties arise by endeavouring to refer the quantities of the four great mineral alkaloids found in the earth's successive strata from the first metamorphic to the last of the tertiary and post tertiary rocks, *to granite as their origin*, and assume that mechanical disintegration, aided by organic action, has been the means of supplying the different elements in their relative proportions to the several strata in which they are found.

To pass from fact to simile. It is as though a salesman had contracted with certain farmers for certain produce—say, for bacon 20 tons, butter 15 tons, lard 5 tons, and cheese 3 tons; and the whole are duly catalogued in the accompanying invoice when delivered, and when the invoice is examined, the order is said to be correctly executed. The goods are presently unpacked and warehoused, and after being carefully weighed, the following discrepancies are found between goods and invoice: bacon, 1 ton; butter, 12 tons; lard, 5 tons; and cheese 30 tons. If for these four edibles the four alkaloids are substituted, and they are placed in the following order, an *approximation* to the truth may be obtained:—In granite: potash 20, magnesia 15, soda 5, and lime 3 parts; and in the sedimentary rocks: potash 1, magnesia 15, soda or sodium 5, and lime 30 parts.

It will be asked—if granite does not supply the lime, from whence does it come, as there is no other source for it, upon the detrital theory? Answer—Reject the theory, hold to the lime, and all the *known* facts of stratified rock composition; and, as it is a very HEAVY subject, weigh it well, and pronounce no opinion *until materials and theory counterpoise each other*. The writer himself will not venture to supply any theory.

But providing no theory can be found to answer, by reason of the difficulty in accounting for the occurrence of one chief element, ought the entire theory of detritus from granite on that account to be rejected, since it is the most plausible source of the constitution of the strata of the earth's crust? Answer—If lime were the only element unaccounted for, still it is so largely distributed, and so important, that any source which fails to supply that one ought to be fatal to the theory.

But there is another element equally widely distributed, and of very considerable amount, and of which granite is almost, *if not entirely*, devoid, viz., carbon—an element which from its mode of combination does not occur to the recollection so readily as lime, and which, in the form of hydro-carbons and carbonates, is almost of universal distribution, and of which the atmosphere, when charged to saturation, could retain but a mere fraction of the amount known to exist. For, though a carboniferous era for the coal measure has been heralded by a very questionable densely-carbonized atmosphere, yet, such an extreme hypothesis being admitted, the utter valuelessness of it is apparent when the amount of carbon contained in lime and magnesia, in the form of carbonates, is borne in mind, which in itself can scarcely be much less, if any, when each are isolated, than that contained in the coal measure. Added to which is the present amount of carbon contained in the atmosphere, and in vegetable and animal products all over the world. And when all these are considered, the failure in granite to supply carbon is most complete.

Having now dismissed the subject of supply and demand between granite and the sedimentary rocks, a friend to the detrital theory would say, "All that has been said about granite sounds very well, but how about the organic remains?" For if there be one fact more certain than another with respect to stratification being the result of the slow wasting and deposition in the sea, or land-locked lakes, of the *débris* of one order or system of creation with its necessarily accompanying soil or rock, slowly consolidated by pressure and mutual attraction of its particles, surely that fact is, that every stratum has its own peculiar organic remains. Thus, Mr. Page, in his "Advanced Text-Book," which gives an excellent summary of the present state of geology, describes the theoretical aspect of the science:—
"By examining, noting, and comparing, as indicated in the preceding paragraphs, the geologist finds that the strata composing the earth's crust can be arranged in series; that one set or series always underlies, and is succeeded by, a different set; *and that each series contains the remains of certain plants and animals not to be found in any other series."*

It is quite true that in the transition between one stratum and the next in order above it, the same organic orders and families will be found, and here and there—as with the Terebratulæ, Productæ, the Ammonites and Encrinites, etc.—different groups and species pass through a long succession of strata, yet the individuals of each stratum have a sufficiently defined marking and proportions to determine which stratum produced them, and if found out of their proper order, receive from the experienced palæontologist the condemnation of being *derived*, and not properly indigenous to that formation.

Without, therefore, any laboured attempt to prove that which geologists already grant, that each stratum contains its own organic remains, within certain geographical areas; or attempting to show that organic remains, though not in a regular gradational order, as a whole observe an ascending or progressive order of organization, it will be only necessary to observe that, from the lower Silurian and Laurentian onwards to the highest of the secondary rocks, or the upper Cretaceous formations, and onwards through the Tertiaries, we have successive forms of algæ, confervæ, mosses, ferns, monocotyledons, as palms and canes, or the order of cycadeoideæ and coniferæ, with the still higher dicotyledons pertaining to the vegetable kingdom. And in the animal kingdom innumerable genera and species now extinct and unrepresented, with genera fully preserved in the existing fauna, but whose precise individual species or family remain distinct from the lowest of the rhizopods or protozoæ to the higher forms of mammalia.

The only general exception to this distinction in the individual species is in the lower forms of protozoic life, where, from the variety and close proximal forms different species undergo from different degrees of pressure, heat, and perhaps light to which they have been subjected, distinct species and genera are occasionally almost indistinguishable from each other, as has been shown in the beautiful monographs by Messrs. Parker and Jones, and the more elaborate work, issued by the Ray Society, upon the Foraminiferæ, by the joint labours of Carpenter, Jones, and Parker. Therefore, in determining their presence or absence as identical species, between one stratum and another, or in past strata and the existing fauna, much care is required, and, as determining agents for particular strata, are, for the most part, not to be depended upon.

From this rapid review of organic conditions, from the lowest of the protozoa to the higher mammalia, let the attention next be directed in reviewing the forces of nature which tended to establish that order of stratification now found in the geological chart of rocks.

The attention is first arrested by the fact that different epochs or intervals of stratification are marked by more violent forces of elevation acting powerfully and somewhat abruptly at one period, and more gradually and continuously during the intermediate periods, as at the present time in Scandinavia, along the coast of the Mediterranean; and in the Pacific along the coast of South America, where elevations or depressions are going on slowly but constantly, and in many other parts.

After the formation of the metamorphic rocks, the elevating forces first appear to have broken forth through the Silurian and Devonian systems in great violence, and to have considerably elevated the submerged rocks, fracturing and dislocating them in bold and unequal proportions; whilst from thence till the Permian system was formed, the same work of increasing elevation of the land was going on, *but in a more gradual and regular manner.* After the Permian system was finished, nature roused up her semi-quiescent forces into more powerful operation, and the Triassic system abounds in bold and rugged mountain scenery, with abrupt fractures and mountain elevations, the evident result of internal forces expanding and raising the crust of the globe above its former level most extensively and violently.

From this period till the completion of the cretaceous system, the same forces, acting more equally and regularly, were increasing the general area of the land above that of the water; and after this system was nearly completed, the tertiary rocks were ushered in by a general and violent action of forces which had been long at work as slow and almost, at parts, imperceptible elevating forces, but were now put forth in all their plenitude and grandeur, as the lofty summits of the Alps, the Pyrenees, and the Andes, with many others, significantly attest; and by such process of elevation, in a short period, the relations between land and sea approximately assimilated themselves to those pertaining to the historic era of man.

Thus from the silurian to the permian, and from the permian to the cretaceous or lower tertiary, the land has been continually gaining in area at the expense of the sea, and the fauna and flora have become more terraceous in their structure and habits, and their organization has more and more rapidly assimilated itself to the necessities and conditions of land habitation.

With this increasing elevation of the land, rivers and fresh-water lakes became more common and extensive; animals of a higher order became more general; whilst the flora of the tertiary series exhibited in elegant contrast the trees and plants of tropical with those of more temperate regions, the soil being carpeted with mosses and grasses analogous to those now existing, where the exogenous and endogenous growths blend and contrast with each other, and frequently succumb and rot through the officious attachment of their less honoured but more tenaciously vital neighbours, the cellularies and their allies.

From, therefore, the foregoing review of the effects of the elevating forces, and the onward progress in the floral and faunal development, it may be inferred that, at least, in the laws of nature or of physics generally, the tertiary formation assimilated itself in every respect to those laws now binding and governing the structure and mechanism of organic and inorganic matter.

That this general inference will be admitted no reasonable doubt can be entertained, since the laborious and masterly work, "The Principles of Geology," by Sir C. Lyell, has been so long and favourably received, not only by British geologists, but by their Continental and Transatlantic brethren. That able author has laboured to show that the entire series of sedimentary rocks need no other forces for their production than those now existing and in constant operation, providing that the birthright of TIME can be sufficiently extended so as to reach back to the limits of their first genesis.

The same subject is somewhat cautiously but very comprehensively summed up by the writer of the article "Geology," in the "English Cyclopædia." He thus writes: —" Successive phases of the aqueous and igneous agencies over the same region appear, either contemporaneously or successively, to have affected all parts of the earth's surface accessible to man; so that everywhere there is proof of great revolutions in the condition of land and sea. Moreover, it appears that to each general system of stratified rocks, indicative of a corresponding great system of physical agencies, peculiar races of plants and animals belong; with new physical conditions new forms of life came on the globe, vanished with those conditions, and gave place to others equally transitory. If, now, we compare the modern survey of nature with any similar work, executed on the same principle, for any one of the earlier epochs, it is certain that the earth has undergone many very extensive revolutions in all that respects its aqueous, igneous, and organic phenomena, before arriving at its present state; it is equally certain that between the epochs of these revolutions the state of the earth was not extremely dissimilar to that which we now behold; yet, because the organic beings preserved in the earth in each of these systems are peculiar to it, and differ from the others and from those that now live, we cannot possibly doubt that the points of difference are numerous, general, and important."

The same writer thus speaks of the tertiary periods:—" *In general*, no contrast can be more complete than that between the secondary and the tertiary stratified rocks; the former retaining so much uniformity of character, even for enormous distances, as to appear like the effect of one determined sequence of general physical agencies; the latter exhibiting an almost boundless variety and relations to the present configuration of the land and sea not be mistaken. *The organic bodies of the secondary strata are obviously and completely distinct from those of the modern land and sea;* but in the tertiary deposits it is the resemblance between fossil and recent kinds of shells, corals, plants, &c., which first arrest the judgment."

If, then, it is granted that the same physical agencies were at work from the commencement of the tertiary period to the present time, and therefore that the known phenomena occurring along our rivers, lakes, shores, and deltas, or oceanic estuaries, were the same then as now, and also that the effect of the trituration of the water on the sides and at the bottoms of rivers was then as it is now, and that at the time of the commencement of the tertiary period the greater part of the earth's surface was covered with secondary rock formations, how is that tertiary formations do not abound in the organic products washed down by, and precipitated from, the waters that have flowed over, say, oolite or Jurassic, the Wealden or chalk formations, or, not to be particular to a shade, say from the coal beds of the carboniferous era, or some of the numerous organic bodies of the silurian age?

It is again repeated, that the laws in operation in the tertiary period being the same as those now in operation, why do we not find, as the effect of trituration and drift, organic bodies transported from the secondary to the tertiary formations with as much ease and frequency as such small objects as boulders are, and now and then small masses of rock? Is it that organic bodies, once petrified and consolidated by pressure, cling with a tenacity stronger than life to their native fatherland?

If such is to be the admitted assumption, how strangely have we ignored our premises. For is there a shore, say a Norfolk crag, a southern chalk cliff, or a northern coal bed, which is washed by the ocean's waves, and the sea border in close proximity does not contain with the *débris* the organic products of that border, whilst the mid-ocean is loaded with organic bodies peculiar to, and distinctive of, the present era?

It is beyond dispute that such a state of things *is now going on*, and every sea deposit has its *derived* organic remains from the shore to which it lies contiguous; no matter what that stratum is in which organic bodies are contained, the contiguity of that rock to a shore occasions the corresponding sea deposit to partake of *derived impurities,* which, so far as letters engraven in stone can attest, most *plainly and silently express the fact that the era which classified organic remains in special compartments has passed for ever.* And should an era follow the present historical era, our successors, by a process of natural selection, being of a higher order than ourselves (providing the SELECTION be not like that of the mare for the male donkey), would find organic remains not only alike in genera and family, but of precise and individual species with those belonging to rocks of old formations, as of chalk, oolite, permian, silurian, &c., &c., mixed and commingled with the flora and fauna of the quarternary or historical formation.

Professor Owen, in his "Palæontology," page 12, thus incidentally alludes to it:—" Most of the fossil genera, and even some of the species, pass through many formations." And concludes by saying—" It has, however, been observed that fossil rhizopods, set free by the disintegration of rocks, *are mingled with the recent shells on every beach;* and Mr. McAndrew has obtained them in this condition from great depths of the mid-channel."

Again, Mr. Page, in speaking of icebergs, says :—" Nay, icebergs have been encountered in the North Sea covered or interstratified with ancient soil, among which were the bones of mammoths and other extinct animals, still further confusing the nature of their deposits by mingling the remains of an existing fauna (reindeer, musk ox, Arctic bear, &c.) with one of a much higher antiquity."

If our logic proceeds from direct facts as they stand revealed in the contents of each stratum, or from general and precisely defined principles, how is it POSSIBLE to maintain that the tertiary periods in their organic remains should be so *distinct from the entire secondary* (saving at their lowest margin with certain chalk formations), and yet that they should be formed from the detritus or washings of numerous preceding strata, and those strata themselves formed in a great measure by *débris* from each other?

To be certified as to the correctness of the conclusions, there is no need for any very lengthened analysis of the contents of each stratum, since, in exact proportion to the increased knowledge of the conditions in which the strata, with their organic bodies are found, will be the increased evidence of the fact that the strata of the earth's crust were not derived from each other, either wholly or in part. Whilst, on the other hand, the explanation or theory of each formation, being derived from previous strata or rock, is so plain and under such simple physical conditions, that the result is *entirely impossible*.

No one has more simply and lucidly described these conditions than Mrs. Somerville ("Physical Geography," 4th edition, 1858):—" Aqueous rocks are all stratified, being sedimentary deposits from water. They originate in the wear of the land by rain, streams, or the waves of the ocean. The *débris* carried by running water are deposited at the bottom of the seas and lakes, where they are consolidated, and then raised up by subterranean forces, again to undergo the same process of destruction, after a lapse of time. By the washing away of the land, the rocks are laid bare ; and as the materials are deposited in different places according to weight, the strata are exceedingly varied, but consist chiefly of arenaceous or sandstone rock, composed of sand, clay, and carbonate of lime."

Taking the above as a correct statement of the detrital theory, the inevitable conclusion arrived at must be that of a *contradiction*, providing it is maintained that every stratum has its own peculiar organic remains, and every stratum derived from such previous order of stratification is, as to entire identity of individual species, perfectly free. Or, in other words, *the derived* must be, in all points of organism, self-created and *independent of the source* from whence it proceeds. And the organic remains being to each stratum peculiar, upon a derived or detrital theory, is self-destructive.

Many other difficulties, upon a detrital theory, might be urged—such as the tertiary basins ascending above and descending below the sea level *in the midst of secondary rocks*, and obtaining successive land and marine remains ; whilst the granite itself, which underlies all rocks, must fracture around the limits of the basin *several successive times*, in such cases, to account for successive formations in a circumscribed area, of both land and marine stratification, with their respective organic remains. But enough has already been said to justify a Doubt relative to the Epochal and Detrital Theory.

INDEX.

ACCELERATION, in attraction constant, in repulsion variable, 60
ACRE an imaginary centre of Dynasties, 260-276
Airy's Explanation of, directly as the mass, 51
Animal tripartite membrane, 6, 82
Animals and Vegetables, distinction between, 9, 141
Antagonism essential to demonstrate force, 38
Apology for brevity, 9
—— for the introduction of Anthropology, 26, 284
ASSOCIATION, a principle of action in present era, 25, 278
Astronomy, by the Ancients, in relation to Longevity, 221
Atoms permeated by two fluids unequally, 55
Atoms have a fixed form, 61
Attraction and Repulsion of unequal Acceleration, 2, 47
—— directly as the mass often misunderstood, 39
——, equal, opposed to Chemical Affinity, 4, 40
—— Similar and Eclectic, 64
Attractive force fixed, Repellent accumulative, 4, 53
Axoidal and Orbital Motion require active forces, 2, 45

BARK of trees in relation to the leaf, 138
Bennett's, Dr. J. H., ideal notion of Vital Force, 37
Black-death, a graft upon Plague, 18
——, by Hecker, 185
—— and Plague a hybrid disease, 189
Blood-letting, its decline and fall, 234
Boyle's Experiment under the air-pump indecisive, 41
Brain and Spinal Chord, metamorphosed muscle, 88, 123
—— a Concentrating Organ, 123
—— centre of a series of limbs, 123
——, its use in relation to the senses, 130
——, Reasons for judging it to be metamorphosed muscle, 135
——, Senses after leaving the, have each a proper sense apparatus, 105

CARBON, its source, 318
Cell differentiation changes with kinds of nutriment, 78
—— comparison between Animals and Vegetables, 79
Cerebellum, supposed compound function, 122
Cervical region withdrawn in fish, and with them no plurality of limbs, 134
Chemical affinity, opposed to equal attraction, 4, 40
Cholera and fungiferous poison, 253
Chronology, Usher's adopted, 204
——, Remarks upon St. Matthew, 205
Circulation, its relation to the state of the heart, 238
Colour, determined by molecular arrangement, 5, 78
——, known as reflected light in organic bodies, 5, 75
——, uniform in the wild condition, but variable under cultivation, 77
Compass, secular variations of, 29, 293
Connective tissue not membrane, 82
Contractile tissue, none in Vegetables, 140
Cross breeds in Cattle and Horses, 258

DIFFERENTIATION and Metamorphosis, 5, 78
Diocletian, Emperor, remarks upon, 164
Disease, change of type, 176
——, its induction and planting, 229

EARTH, the, being once a molten mass, rejected, 310
Earth's internal magnetic mass, 292
—— Stratification, its origin, 297
—— sudden thermal changes, 291
—— surface and interior, suffered at some time a great physical change, 294-297
Earthquakes and Volcanoes not causes of Epidemics, 287
Electricity, its distribution to the heart, and the muscles terminating arteries, 92-98
Endemic conditions the result of civilisation, 10, 143
—— conditions of a hereditary origin, 145
—— and Epidemic diseases in relation to Vital Force, 146
Epidemic Eras, their duration, 16
—— Eras, an outline of their course, 18
——, present Era, 23, 299
—— in present Era, Heart most affected, 23, 235
—— Eras last about six hundred and forty years, 161, 177
—— Eras in relation to the Rise and Fall of Empires, 25, 260
——, the Chronographic, 217

Epidemics, Comparison with the Rise and Fall of Empires, 283
——, their Supposed Causes, 18, 286
——, General Summary, 294
——, General Observations, 10, 152
——, ISOLATION a principle or law in, 13, 150
——, their progress, 160
Erectile tissue in relation to mammary membrane, 93
Ethical code equal for general civilization, 278
Evergency a principle of Mechanism, 111
—— and Invergency, 136
Exuviæ as producers of lime, 308

FISH, axis of muscular motion in, 108
Floral changes, 259
Food affecting development, 257
Force undemonstrable without Antagonism, 2, 38
——, or influence from Sun to Earth, probably restored, 74
Forces in Nature which regulate Stratification, 321
Fracture, case of compound, Sympathy between bone and integument, 108

GEOLOGY.—Epochal theory untenable, 29
    The Earth never a Molten Mass, 35, 310
    Four points considered as adverse to the Epochal Theory, 33
    Granite not the Source of all our Strata, 35, 299
    ——, Composition of, etc., 301
    ——, Comparison of Stratified rocks and Granite, 304, 317
    Organic remains peculiar to each Stratum adverse to Detrital Theory, 35, 319
    Outline of duration of Epochs, 31, 299
    Special adaptation of the Earth's crust for new orders of organized products, 32
Geology, Pamphlet upon, Anonymous (1864), 33
——, Tertiary conditions of fauna and flora, 322
Gibbon, Extract upon the Levant Plague, 181
Gravitation accounts for Orbital Motion not Axoidal, 43
—— the result of Attraction, 38
Granite fracture about the *same* area to great depths, many times, 327
Great Pyramid in relation to Longevity, 222
——, for why it was built, 225
Growth of Animals and Vegetables in present Epidemic Era, 254

HEALTH in relation to moisture, etc., 13
Heart's action feeble in present Epidemic Era, 235
—— and Lungs, sympathy between them, 253
Heat a motor power, especially in germ development, 71
—— as a factor of disease, 288
——, its depth *within* the Earth, 312
—— a Repellent force between Atoms, 70
Herschel, Sir J., upon authentic weights and measures, 227

IMPRESSIONS on each kind of germ, and impetus by the tangent compared, 68
Inertia, passiveness of, 44

JOINTS, not muscles, the correct location for taking cognizance of resistance and weight, 117

LANGUAGE of Science inexact, 41
Leaf in relation to the density of Wood, 140
—— in relation to the Bark of Wood, 138
Leprosy an early infectious disease, 17
——, did Pompey's Army bring it from Egypt? 168
——, when it first reached England, 162
—— appeared in Europe, 167
——, its infectious and hereditary conditions, 172
—— in the days of Moses not hereditary, 173
——, *true*, not mentioned in history from 750 to 64 B.C., except in Egypt, 210
——, its origin by Manetho, 211
Light, undulatory theory of, some things which favour it, 76
—— in relation to colour, 5, 75
Lime in sea-water, 306
——, proportion of, between stratified rock and granite, 305
——, Supply from Eruptive rocks, 309, 315
Lung affection in Cattle, 248
Lungs, Chronic congestion and its diagnosis, 241

MAGNETISM and Attraction, 149
——, Electro, Earth's poles N. and S., 292
Mammalia, homologies of limbs and pelvis in, 115
Matter permeated by Antagonistic fluids, 56
Mechanism, principles of, in Animal and Vegetable Kingdom, 8, 135, 142
——, principles and laws of Nature in union with each other, 142

Membrane, Animal Tripartite, defined as contractile, serous and mucous, 7, 80
Membranes, Tripartite, number in Mammalia, 81
——, Alimentary, 83
——, Circulatory, 90
——, Ganglionic, 92
——, Genito-urinary, 84
——, Lacteo-lymphatic, 91
——, Mammary, 90
Membrane, Nasal, probably a distinct, 84
——, true animal, essentially quadruple, 103
——, true animal, essentially segmented, 104
Metamorphosis and differentiation, 5, 78
Morphological Changes, the outcome of Creative Will on the germ, 72
—— completed in the cycle of a year, but lunar changes affect lower forms the most, 73
Mountains, base and elevation of, opposed to great depth in the heat of the Earth, 313
Muscle, striped, involuntary under certain conditions, 121
Muscular Motion, axis of in Fishes, 112
—— in Birds, 113
—— in Mammalia, 114

NATURE, her laws and principles of Mechanism harmonise with each other, 142
——, her law the same in past eras of Geology as in the present, 322
——, Precursion the great law of, 4, 69

OERSTED upon Magnetism, 149
Organic and Inorganic Kingdoms are under the same laws, 1
Organic remains distinct in each stratum, 35, 319
—— in relation to epochs, 319
—— not derived from one stratum and supplied to another, 326
——, why in strata by detritus they do not intermix, save those of present era, 325
Organization, mechanism of, and the laws of nature, their distinctiveness, 29
—— the result of Will, and under law, 68

PARTURITION, an excretory function, 87
——, lime a stimulant in, 88
Patriotism, a word misapplied, 260
Plague, Athenian, a hybrid disease, 19, 199

Plague, Athenian, by Thucydides, 193
—— and Black-death Hybrid, 179
——, extract from Gibbon, 181
——, Small-pox, and Leprosy widespread Epidemics, 15, 160
——, and Small-pox North and South of the Mediterranean, and afterwards spread to Mid and Northern Europe, 16, 179
Poisons in Epidemic, Animal and Vegetable, 230
Polypifera and Sponges have a Contractile Membrane, 81
PRECURSION, the great law of Nature, 4, 69, 72
Principles of Mechanism in the Animal and Vegetable Kingdoms, 8, 135, 142
PROPOSITIONS aiming to unify the laws of Motion and Chemical Affinity, 54
Pyramid, the Great, in relation to longevity, 222
——, why was it built, 225

RACES in Africa, remarks upon, 281
—— in Asia, remarks upon, 282
—— of Mankind, their distinctiveness, 279
Repulsion, an accumulative force, 50
—— in proportion to the Mass, 53
Résumé, 27

SEA-WATER, its composition, 306
——, Bischof's assumption, 307
Seasons, wet, windy, hot, or cold, not causes of Epidemics, 152
Sense apparatus, each has its own, 102
—— Hæmal, 106
—— Subdermal, for Force, 107
—— of Touch, 118
—— of Force, essential to realise resistance, 117
—— of Smell, its absence and withdrawal of cervical region, 133
—— Telluric or travelling, 126
Senses, educators of each other, 125
——, Seven special, 8
—— of Smell, Sight, and Hearing elongated limbs, abridged by Mechanism, 8, 124
——, Three Somatic, of Want or Hæmal, of Force or Weight, and of Touch, 99
——, division of, into active and passive, 128
——, into Somatic and Paraitic, 129
—— Paraitic, outlines of apparatuses, 130
—— of Touch and Force in relation to integumentary distribution, 119

Senses of Touch and Force use the neural arches in common, 120
Small-pox, remarks upon, by Dr. F. Adams, 165
——, and Plague, their spread after 1177 A.D., 166
Spinal Chord and Brain metamorphosed muscle, 8, 88, 123
Spirits, Remarks upon their use, 238
Sun, direction of his rays in relation to heat and vital phenomena, 147
——, his force, or influence upon the Earth, probably restored, 74
——, his heat, regulated by the internal condition of the earth, 147
——, a non-producer of Epidemics, 148
——, Spots on the, in relation to Epidemics, *nil*, 150
——, a Vitalizing agent, 11, 146

TANGENTAL Force, its use, 2
—— never expended, 42
—— only gives direction to motion, 49
Temperature, compared for 3,000 years, 27, 288
——, the Earth's internal condition regulates heat on the surface, 28, 147
——, as by Wind, Monsoons, and Frosts, 289

VARIÆ, 1-328
Vital conditions affected by internal magnetism, 295
—— Physics, 1, 37
Volcanoes and Earthquakes not causes of Epidemics, 287

www.ingramcontent.com/pod-product-compliance
Lightning Source LLC
Chambersburg PA
CBHW032047220426
43664CB00008B/904